W9-BGN-684

Developing Resource-Informed Strategic Assessments and Recommendations

Paul K. Davis, Stuart E. Johnson, Duncan Long,
David C. Gompert

Prepared for the Joint Staff
Approved for public release; distribution unlimited

 NATIONAL DEFENSE RESEARCH INSTITUTE

The research described in this report was sponsored by the Joint Staff. The research was conducted in the RAND National Defense Research Institute, a federally funded research and development center sponsored by the Office of the Secretary of Defense, the Joint Staff, the Unified Combatant Commands, the Department of the Navy, the Marine Corps, the defense agencies, and the defense Intelligence Community under Contract W74V8H-06-C-0002.

Library of Congress Cataloging-in-Publication Data

Developing resource-informed strategic assessments and recommendations / Paul K. Davis ... [et al.].
      p. cm.
   Includes bibliographical references.
   ISBN 978-0-8330-4502-7 (pbk. : alk. paper)
   1. United States—Strategic aspects. 2. Military planning—United States.
3. Resource allocation—United States. 4. National security—United States—21st century. 5. Strategy. I. Davis, Paul K., 1943–

UA23.D287 2008
355'.033573—dc22

                                                      2008035694

Cover illustration by ImageZoo courtesy of Media Bakery.

The RAND Corporation is a nonprofit research organization providing objective analysis and effective solutions that address the challenges facing the public and private sectors around the world. RAND's publications do not necessarily reflect the opinions of its research clients and sponsors.

RAND® is a registered trademark.

© Copyright 2008 RAND Corporation

All rights reserved. No part of this book may be reproduced in any form by any electronic or mechanical means (including photocopying, recording, or information storage and retrieval) without permission in writing from RAND.

Published 2008 by the RAND Corporation
1776 Main Street, P.O. Box 2138, Santa Monica, CA 90407-2138
1200 South Hayes Street, Arlington, VA 22202-5050
4570 Fifth Avenue, Suite 600, Pittsburgh, PA 15213-2665
RAND URL: http://www.rand.org/
To order RAND documents or to obtain additional information, contact
Distribution Services: Telephone: (310) 451-7002;
Fax: (310) 451-6915; Email: order@rand.org

Preface

This monograph reports the results of a project to provide the Joint Staff's Vice Director for Force Structure, Resources, and Assessment (J-8) with methods, desk-top tools, and initial data to help the Chairman of the Joint Chiefs of Staff develop resource-informed assessments and recommendations for the Secretary of Defense on national military strategy.

The project was requested and sponsored by the Vice Director, J-8, MG Michael Vane (USA) and was completed under his successor, MG William Troy (USA). It was co-sponsored by the Office of Force Transformation in the Office of the Secretary of Defense (OSD) for Policy and by the Office of the Secretary of Defense for Program Analysis and Evaluation. The monograph should be of interest primarily to those senior leaders and their staffs—military and civilian—who are involved in the Department of Defense's (DoD's) strategic planning. It should also be of interest to strategic planners in other government agencies. Comments and suggestions are welcome and should be addressed to the project leader, Paul K. Davis, in Santa Monica, California (email: pdavis@rand.org; telephone: 310-451-6912).

This research was sponsored by the Joint Staff and was conducted within the International Security and Defense Policy Center of the RAND National Defense Research Institute, a federally funded research and development center sponsored by the Office of the Secretary of Defense, the Joint Staff, the Unified Combatant Commands, the Department of the Navy, the Marine Corps, the defense agencies, and the defense Intelligence Community.

For more information on RAND's International Security and Defense Policy Center, contact the Director, James Dobbins. He can be reached by email at James_Dobbins@rand.org; by phone at 703-413-1100, extension 5134; or by mail at the RAND Corporation, 1200 S. Hayes Street, Arlington, VA 22202. More information about RAND is available at www.rand.org.

Contents

Figures

Tables

Summary

Background

The United States will soon be conducting another major review of national-security strategy. It will be the responsibility of the Chairman of the Joint Chiefs of Staff (JCS) to provide resource-informed assessments and recommendations to the Secretary of Defense (SecDef) and the President. This monograph illustrates newly developed methods and tools to support the chairman's efforts. We sought a way to compare strategies that would integrate expectations about effectiveness, risks, and resource implications. Such an approach would tie into the Department of Defense's themes of capabilities-based planning, risk management, and portfolio analysis. To permit timely responses to senior-leader guidance, questions, and feedback, we put a premium on relatively simple methods.

Approach

In developing a strategic planning approach, we drew on the past history of defense planning and strategic planning in large business organizations. A central concept is viewing issues through what the business world calls an *operating-unit perspective*. We consider DoD's operating units to be the combatant commands (COCOMs) plus a virtual "National Command" associated with the Secretary of Defense and supported by the chairman.

Figure S.1 sketches the approach. Given a set of alternative national strategies, the approach does the following for each strategy in turn: characterizes its premise, goals, and approach; characterizes the operating-units' objectives; characterizes capability needs and implications for forces and force capabilities; and estimates costs and other resource implications. The last of these includes ascribing costs to the capabilities added to (or taken from) each COCOM, even though those costs are budgeted through the services. This is analogous to the use of "transfer costs" in business (i.e., billing operating units for what their suppliers provide, even though the suppliers are actually tasked and paid directly). The intent is to enable senior decisionmakers to clearly see the link between strategic changes and resource implications and to enable operating units to lobby effectively for changes when they are troubled by disconnects among responsibilities, authorities, and resources.

At a more subtle level, we sought both to further progress in global thinking and military jointness and to honor what we see as the natural partnership between joint and service planners. U.S. military services are budgeted separately by Congress to recruit, train, and equip. However, they are not mere "suppliers" akin to those of a commercial marketplace. They are deeply involved in strategic planning, research, innovation, and experimentation. It is the services that actually develop the capabilities that joint commanders employ. They do this with future joint contexts strongly in mind. Our approach does not contemplate changes in the way programming and budgeting are accomplished technically—with nearly all funds flowing through the services and defense agencies.

The next part of the approach (Figure S.1, bottom) is an integrated assessment using a portfolio-analysis structure assessing strategies for likely effectiveness, risks, and costs. The assessments are for each COCOM separately and then from a national perspective. As suggested by Figure S.1, the process is iterative, because national leaders must reconcile what they desire with what can reasonably be obtained. Strategic planning is neither top-down nor bottom-up; rather, it has elements of both.

Figure S.1
Overview of the Analytic Approach

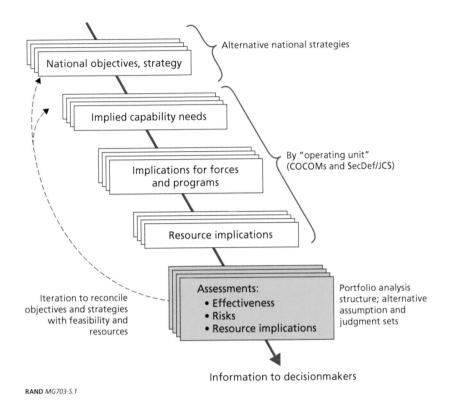

RAND MG703-S.1

Comparison of Illustrative Strategies

It was not our project's purpose to conceive alternative strategies, but we needed concrete options to develop and illustrate the approach. Thus, we developed three alternative strategies that are intended to be topical, provocative, and illustrative—starting points for subsequent work. All are defined relative to an Analytical Baseline comparable to the substantial current U.S. force structure and program, but without the current program's increase in ground forces or heavy involvement in Afghanistan and Iraq. The alternatives, then, are

1. *Direct GWOT/COIN.* This strategy focuses on the global war on terrorism (GWOT) and counterinsurgency (COIN) against violent Islamists acting against U.S. interests. Intended to reflect aspects of actual U.S. strategy earlier in the decade, it depends on substantial direct involvement of U.S. forces for COIN efforts, primarily in the Greater Middle East. The strategy is motivated most strongly by near- and mid-term considerations, although it anticipates a "long war." It gives lesser priority to the Far East.

2. *Build Local, Defend Global.* This strategy also focuses on the Greater Middle East and violent Islamism but is philosophically different. It envisions extensive assistance to locals, building up their COIN capabilities and establishing good partnerships. This strategy would emphasize special operations forces (SOF), maritime operations, and training teams but avoid use of regular ground forces. It would include much more foreign assistance, which would be managed largely by the State Department.

3. *Respond to Rising China.* This strategy proceeds from the premise that, despite Middle Eastern problems, the rise of China is the most important reality around which to design strategy. It seeks to avoid a vacuum in the Western Pacific and East Asia—i.e., to compete effectively with China so as to deter or dissuade actions contrary to long-term U.S. interests, but without provocation or the expectation of an arms race. It puts relatively more emphasis on the long term than do the other strategies. Its approach to the threat of violent Islamism is philosophically similar to that of the Build Local, Defend Global strategy, but with drastically less funding and commitment.

All strategies were forced to adhere to some principles. All should recognize worldwide U.S. interests and concerns, including uncertainties that are both broad and deep. A strategy focused on the Middle East would need to maintain capabilities in the U.S. Pacific Command (PACOM) and elsewhere; also, each strategy had to include various hedges—i.e., had to plan for strategic and operational adaptiveness. This was in contrast to allowing strategies that would "bet the farm" on a particular view of the future. This said, each strategy takes risks differently.

Characterizing the Strategies

Figures S.2–S.5 summarize the strategies' implications for force shifts and programs relative to an Analytic Baseline (Figure S.2), which projects DoD spending of $10.2 trillion dollars over 20 years (not counting supplementals). This Analytic Baseline is similar to today's posture and program, but without the scheduled increase in conventional ground forces or the intense ongoing counterinsurgency activities. That is, it assumes substantially fewer U.S. forces in Iraq and Afghanistan than is the case today but assumes that the Middle East is a top priority. The Analytic Baseline, then, is not the current reality, but rather something against which to compare, arguably comparable to what was implied by strategy at the beginning of the decade (described in Rumsfeld, 2001).

Our characterization focuses only on major units and—in a departure from common practice—associates force units with their "usual" COCOM, even though this is somewhat artificial, since the vast majority of the units are potentially available for deployment to any COCOM. This association was necessary as part of the operating-unit orientation.

These major units account for about $3.2 trillion in DoD expenses over 20 years, leaving $7 trillion unrepresented in the Analytic Baseline. This $7 trillion, which accounts for everything from base infrastructure to support units, is constant across the strategies. Further, only a comparatively small portion of the $3.2 trillion accounted for in the stated baseline is altered in any way. Some cuts and reallocations are made, but all strategies are founded on an already substantial body of resources.

The Direct GWOT/COIN strategy makes changes relative to the Analytic Baseline as summarized in Figure S.3. It adds numerous regular ground forces and special training units; it also includes some security and foreign assistance. Central Command (CENTCOM) gains

Figure S.2
Analytical Baseline, 2009–2028 (2009 $B)

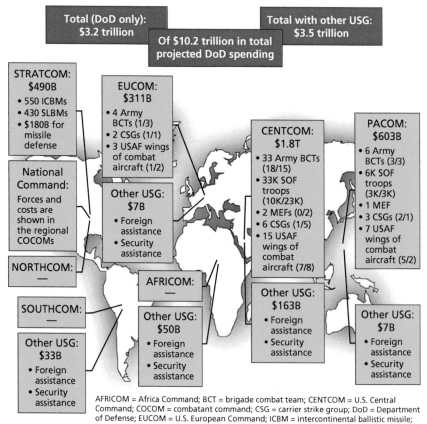

Total (DoD only): $3.2 trillion

Total with other USG: $3.5 trillion

Of $10.2 trillion in total projected DoD spending

STRATCOM: $490B
- 550 ICBMs
- 430 SLBMs
- $180B for missile defense

National Command: Forces and costs are shown in the regional COCOMs

NORTHCOM: —

SOUTHCOM: —

Other USG: $33B
- Foreign assistance
- Security assistance

EUCOM: $311B
- 4 Army BCTs (1/3)
- 2 CSGs (1/1)
- 3 USAF wings of combat aircraft (1/2)

Other USG: $7B
- Foreign assistance
- Security assistance

AFRICOM: —

Other USG: $50B
- Foreign assistance
- Security assistance

CENTCOM: $1.8T
- 33 Army BCTs (18/15)
- 33K SOF troops (10K/23K)
- 2 MEFs (0/2)
- 6 CSGs (1/5)
- 15 USAF wings of combat aircraft (7/8)

Other USG: $163B
- Foreign assistance
- Security assistance

PACOM: $603B
- 6 Army BCTs (3/3)
- 6K SOF troops (3K/3K)
- 1 MEF
- 3 CSGs (2/1)
- 7 USAF wings of combat aircraft (5/2)

Other USG: $7B
- Foreign assistance
- Security assistance

AFRICOM = Africa Command; BCT = brigade combat team; CENTCOM = U.S. Central Command; COCOM = combatant command; CSG = carrier strike group; DoD = Department of Defense; EUCOM = U.S. European Command; ICBM = intercontinental ballistic missile; MEF = Marine Expeditionary Force; NORTHCOM = Northern Command; PACOM = U.S. Pacific Command; SLBM = sea-launched ballistic missile; SOF = special operations forces; SOUTHCOM = Southern Command; STRATCOM = U.S. Strategic Command; USAF = U.S. Air Force; USG = U.S. government.

NOTES: Notation (1/3), e.g., means that 1 unit is committed to a COCOM and 3 are held in National Command.

RAND MG703-S.2

the great majority of the new resources. The total resource implications of the strategy are to increase expenditures by $248B for the DoD and by $302B for the U.S. government overall over 20 years.

The Build Local, Defend Global strategy deemphasizes ground forces relative to the baseline. It reduces ground forces earmarked for CENTCOM and PACOM by two and three brigade combat teams

Figure S.3
Force Shifts and Program Initiatives in the Direct GWOT/COIN Strategy

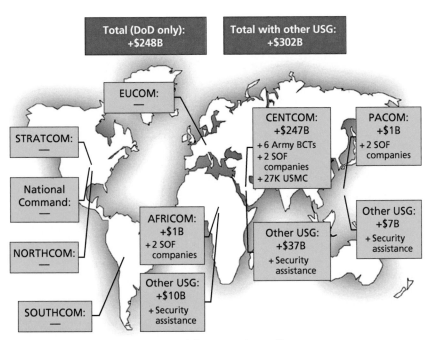

NOTE: Numbers may not add to totals because of rounding.
RAND *MG703-S.3*

(BCTs), respectively, moving three of these to the National Command as an uncommitted and unoriented strategic reserve of active forces.

The Army converts a BCT-equivalent of its remaining CENTCOM forces into military trainers and advisors, with most remaining in CENTCOM and the rest available for global deployments. The strategy also adds capabilities for training and units for intelligence, surveillance, and reconnaissance (ISR). So-called "green water squadrons"—units with small but capable ships—are added to AFRICOM, CENTCOM, SOUTHCOM, and PACOM to foster maritime security partnerships and improve littoral capabilities. The strategy also adds to the National Command additional SOF and ISR units and begins procurement of long-range reconnaissance and strike aircraft.

The total resource implications for the Build Local, Defend Global strategy are to decrease DoD expenditures by $28B (FY 2009$) over

Figure S.4
Force Shifts and Program Initiatives in the Build Local, Defend Global Strategy

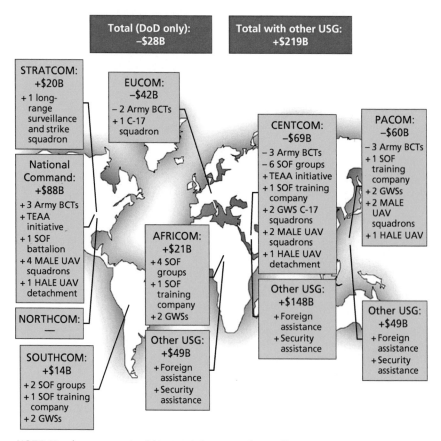

NOTE: Numbers may not add to totals because of rounding.
RAND *MG703-S.4*

20 years, but increase overall U.S. government (USG) expenditures by $219B (FY 2009$) over the same period.

Figure S.5 depicts highlights of the Respond to Rising China strategy. Over the course of the 20 years, this strategy adds significant naval forces and some ISR units to PACOM. In addition, the assets associated with STRATCOM are increased with long-range bombers and missiles, ISR, and improvements in theater and national

**Figure S.5
Force Shifts and Program Initiatives for the Respond to Rising China
Strategy**

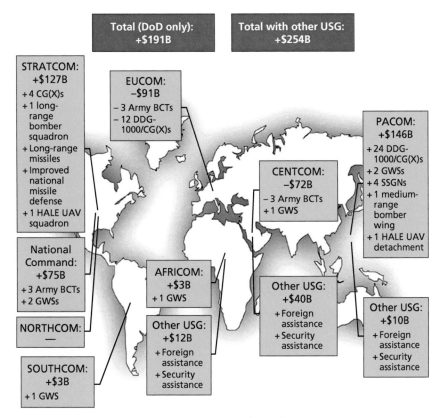

NOTE: Numbers may not add to totals because of rounding.
RAND MG703-S.5

missile defense. Ground forces assigned or earmarked to CENTCOM
are reduced, with some units moving to the National Command.
Green water squadrons are created for CENTCOM, AFRICOM, and
SOUTHCOM. Foreign assistance is increased, but only about 20
percent as much as in the Build Local, Defense Global strategy. One
reason for the increased assistance (beyond the problem of violent Islamism) is to address expanding Chinese influence in Africa.

The total resource implications for the Respond to Rising China
strategy are to increase DoD expenditures by $191B (FY 2009$) over

20 years and overall USG expenditures by $258B (FY 2009$) over the same period.

Although each reader might define programs for the several strategies somewhat differently, the choices we made illustrate differing emphases. All are global strategies, and all make only marginal changes to the fulsome baseline. Thus, most programmed capabilities are not highlighted explicitly (e.g., procurement of F-22s and F-35s, the current version of the program for ballistic-missile defense, or continuation of the Army's Future Combat System program).

The Economics of Strategy in Different "Currencies"

The Different Currencies

Figure S.5 summarizes 20-year costs in constant dollars, but our costing includes nominal Future Year Defense Plan (FYDP) and constant-dollar calculations, 20-year figures based on life-cycle considerations, expenses to the U.S. government as a whole rather than just to the DoD, and the net present value (NPV) of future obligations being made under the strategies. Further, we concluded—in a break from past practice—that responsible costing must consider the extraordinary expense of war or other intensive military operations, which are not typically included in defense planning. These include funds for deployments, combat pay, and recapitalization of equipment worn out by operations, for example. Specialized reports are also needed to show, for example, the implications of a given strategy for each of the military services. *None* of these different expressions of cost is uniquely right, and all are necessary. Appendix C describes a simple tool that we used to generate reports quickly on demand.

It is especially important to consider all costs to the U.S. government when providing resource-informed assessments and recommendations to the Secretary of Defense and the President because the strategies are, ultimately, "national." The Build Local, Defend Global strategy would actually cost the DoD less than the baseline (Figure S.4), but it posits a large increase in foreign and security assistance (mostly through the State Department and its Agency for International

Development, USAID), without which the strategy would be undercut to a degree that is hard to estimate.

Uncertainties in Costing

The various cost calculations are not cut-and-dried. Strategic planning is arguably best done in net present value terms, which makes the point in Figure S.6. This shows that the relative cost in NPV terms of the three strategies is quite different depending on whether one uses a 3 percent or a 7 percent real discount rate (the set of values suggested by the Office of Management and Budget) and on whether one considers all future obligations in such calculations (right side, shown as "indefinite horizon") or only those for the next 20 years. Strategies can be made to seem more or less expensive, even on a relative basis, depending on how their costs are calculated.

Cost of Extraordinary Operations

We have left the most important cost issue until last. The foregoing discussion—consistent with long-standing tradition in U.S. force

Figure S.6
Comparison of Core Costs of Strategy as Function of Discount Rate and Horizon

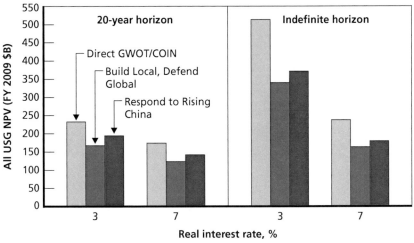

planning—has been about "core expenses" related to force posture, training, and routine operations. That costing does not include the expense of wars or other intensive operations such as occurred in the first Gulf war, the conflicts in the Balkans, or the ongoing campaigns in Afghanistan and Iraq. In the traditional view, such extraordinary expenses could come about under any strategy and would be paid for as a separate matter (i.e., with budget supplementals). When considering its grand strategy for the years ahead, however, the United States must recognize that some strategies are more likely to involve such operations than others. The Direct GWOT/COIN strategy (which is more like today's operations than the other strategies) virtually implies that such operations will occur: Proactive direct involvement is a tenet of the strategy. Therefore, it is legitimate to include those costs in the estimates of the cost of strategy. Figure S.7 does so, using a range of estimates that are 50 to 100 percent of what might be estimated based on activities of the last half-dozen years. The primary observation is this:

Figure S.7
Cost Comparisons Including "Extraordinary" Costs of Operations

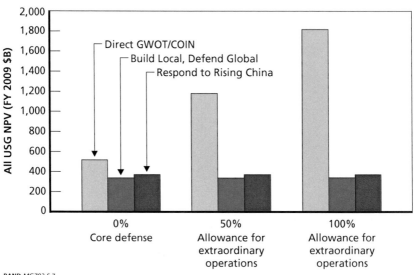

- The relative costs of strategy are dominated by the "extraordinary" costs of actual operations with ground forces.

The issue is debatable, of course. Proponents of the Direct GWOT/COIN strategy might argue that only with such a strategy could the United States expect to avoid an even larger and more costly hot war. Proponents of the other strategies would disagree.

An Integrated Assessment Using Portfolio Analysis

In evaluating strategies, the real issue is whether the *combination* of a strategy's expected effectiveness, risks, and costs makes it attractive. This is the kind of issue for which portfolio-management methods can be useful—i.e., methods for investing in *mixes* of capabilities to deal with multiple and somewhat contradictory objectives while working within a budget. For the current study, we extended and adapted RAND's portfolio analysis tool (PAT), which can be quite helpful in structuring analysis, whether of alternative high-level strategies or of alternative strategies to accomplish something more pointed, such as ballistic-missile defense or global-strike capability. The principal aim is to provide an integrated view of the whole, but one that allows delving into details as necessary to question assumptions, identify alternatives, and otherwise reason about the choices.

Effectiveness

Figure S.8 shows our high-level summary display of effectiveness results, with costs added in the last column. As usual in scorecards, the colors red, orange, yellow, light green, and green correspond to results that are very bad, bad, marginal, good, and very good, respectively. We see, for example, that the expected consequences for PACOM in the Direct GWOT/COIN strategy base case are said to be poor. The assessments in this figure are merely our subjective determinations for this illustrative study but could be based on results of in-depth analysis and senior-leader judgments. The costs given in the right column are all in NPV terms, assuming a 3 percent real discount rate and an infinite

Figure S.8
Top-Level Comparison of Strategies' Effectivenesses

Options	Measures of Option Goodness, by COCOM and Risk											USG Costs (NPV, 3%, $B)
Measure	PACOM	CENTCOM	NORTHCOM	EUCOM	SOUTHCOM	AFRICOM	STRATCOM	SOCOM	National Command	Simultaneous war risk	Overall risk	
Investment options	Detail	Detail	Detail	Detail	Detail	Detail	Detail	Detail	Detail	Detail	Detail	
Analytic Baseline	O	Y	LG	LG	LG	O	O	Y	Y	O	O	0
Direct GWOT/COIN	O	LG	LG	LG	LG	Y	O	LG	Y	R	R	520
Build Local, Defend Global	Y	Y	LG	LG	LG	LG	O	LG	LG	O	Y	340
Respond to Rising China	G	O	LG	LG	LG	Y	G	Y	Y	Y	O	370

NOTES: The costs for Direct GWOT/COIN do not include "extraordinary expenses" associated with intensive operations. These dominate, if included. Letters are abbreviations: R, O, Y, LG, and G for red, orange, yellow, light green, and green.
RAND MG703-S.8

time horizon, and accounting for all costs to the U.S. government (in billions of dollars).

We see that—despite the intent that all of the strategies be sensible—all of them have significant shortcomings as indicated by red or orange cells. Strategic planning is iterative, however. Each strategy's shortcomings could be mitigated with some additional features (albeit at some expense).

As discussed in Chapter Five, our analysis was structured so that staff conducting a study, or senior leaders reviewing it, can zoom (drill-down) into detail, as shown schematically in Figure S.9.

At the lowest level of Figure S.9, for example, the assessments relate to the expected results of future wars used as test cases. The example is for PACOM and assumes that using two test cases for Taiwan and two for Korea would prove adequate; the A and B test cases might corre-spond to a relatively nominal scenario and a particularly difficult one. J-8 and OSD's Program Analysis and Evaluation (PA&E) are heavily involved in simulation-based campaign analysis as part of the Depart-ment's Analytic Agenda. The groups involved could readily identify appropriate summary test cases to be used to feed the portfolio analy-sis. Analysis could also characterize the operational risks (e.g., risks of

Figure S.9
Zoom (Drill-Down) Schematic for Visual Explanation of Scorecard Results

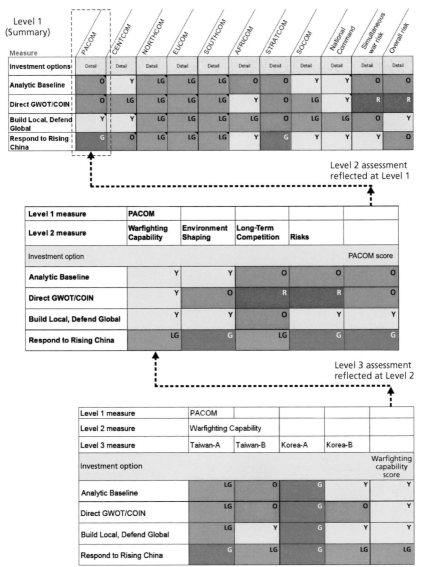

RAND MG703-S.9

even worse actual scenarios or of underestimating adversary capabilities and deviousness).

Other measures are less amenable to simulation-based analysis, but other kinds of studies, perhaps conducted or sponsored by J-5 and OSD (Policy), could characterize the expected consequences of the strategies for long-term competition and environment shaping (key elements of the second-level assessment, as indicated in the middle of Figure S.9). However, it would be for an analytical group to assure that results were scaled in a way commensurate with the more quantitative measures used in the portfolio analysis.

An attractive feature of this analytic approach is that it lends itself well to either deliberate analysis over many months or rapid-paced analysis. Strategic analysis in an iterative environment could be done with senior analysts and officers making reasoned judgments at a high level of the portfolio structure (the middle level of Figure S.9). Assuming expertise (the result of prior analysis and experience), structure, ruthless objectivity, and candor, such work might be better, not merely faster, than would be possible in a deliberate process with committees, logrolling, and the potential for missing the point by sticking too exclusively to on-the-shelf detailed work.

Risk Management

Risk management is a major goal of sensible strategic planning and a special concern of the chairman and secretary. In this study, we developed a fairly rich depiction of the various risks associated with the strategies. Some of these are "accepted" aspects of a strategy: If one has limited resources, giving priority to one demand means running some risks with regard to another. Other risks are less evident but crucial. These include the risk that "best estimates" of a strategy's effectiveness in a particular COCOM's area of responsibility will be completely wrong. For example, a strategy calling for intensive use of U.S. ground forces and special operations forces in the Muslim world might prove counterproductive. Such issues are discussed in Chapter Five and Appendix E.

Exploratory Analysis Under Uncertainty

A fundamental problem in assessing effectiveness and risk is massive uncertainty. Analysis results can differ substantially depending on whether the assessor is oriented more heavily toward one region or another or toward near-term or long-term problems. Results also change substantially depending on the assessor's approach to global risk, as manifested by concern about the possibility of simultaneous conflicts. Such issues cannot be resolved by committee, by proclamation of standard planning scenarios, or by any other simple expedient. It is in the very nature of strategic decisionmaking to view the problem from these different perspectives, recognizing that balancing these perspectives will often drive choices.

Consider this illustration: Suppose that we wish to compute "cost effectiveness." Usually, this means dividing a single composite measure of effectiveness by a single measure of cost. Alternatively, one can plot the composite effectiveness versus cost. It is easy to do such calculations using PAT, but it is also dangerous. Figure S.10 illustrates how the cost-effectiveness comparisons of strategies differ for what we refer to as CENTCOM-leaning, PACOM-leaning, JCS-conservative, and JCS-optimistic perspectives. These differ in how much relative weight is given to the individual COCOMs, how much credence is given to more stressful warfighting scenarios, and the assessment of the probable effectiveness of "direct intervention" in the Middle East (see Chapter Five for details). The figure also shows the effect of considering the extraordinary costs of operations (represented by a horizontal line for the cost of Direct GWOT/COIN). For the particular analysis we did, the Direct GWOT/COIN strategy has the highest composite effectiveness only in the CENTCOM-leaning perspective, and then only slightly. In all other perspectives, the Build Local, Defend Global strategy is superior. A core conclusion here is that

Exploratory analysis under uncertainty is fundamental to the support of strategic planning: Results based on "best estimate" assumption sets and the "predominant" perspective will often be seriously misleading.

Figure S.10
Effect of Perspectives on Cost-Benefit Calculations Using USG Costs

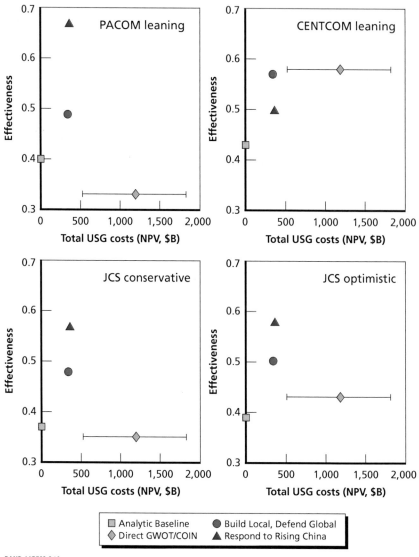

Much progress has been made in learning how to conduct exploratory analysis in recent years, but doing so within portfolio analysis poses special challenges.

Iterating Strategies to Better "Balance" the Portfolio

Although none of our strategies were single-mindedly focused on a single region or objective, there were distinct differences among them. They would lead us to expect such natural questions from the chairman and secretary as "What would it take to amend the such-and-such strategy so that it would do better across the board?" Iteration would then occur. In the extreme, the United States could just "buy everything," but, in practice, choices must be made. The meta strategy (i.e., the strategy of choosing a strategy) should be to achieve flexibility, adaptiveness, and robustness (FARness) of capabilities. This is in contrast to "overoptimizing" for the currently popular prediction of the future and future crises. Supporting analysis, then, should help leaders identify uncertainties and risks and find ways to at least mitigate them inexpensively while responding appropriately to national priorities.

Another type of iteration would involve asking "How much is enough?" More foreign aid and security assistance may well be needed, but would the large investments suggested in the Build Local, Defend Global strategy really pay their way? Could they be trimmed, at least until there was evidence that such investments were successful?

Next Steps for Applications and Research

Our project was a pilot effort intended to illustrate ideas concretely. A number of next steps are possible—both substantively (as in developing and assessing "real" strategies) and methodologically. Chapter Six includes suggestions on the matter and notes that such work would likely be cross-cutting—of interest, for example, to the Joint Staff's J-5 and J-8 and to OSD's PA&E and Acquisition, Technology, and Logistics (AT&L).

Acknowledgments

We benefited from excellent and critical reviews from RAND colleague David Ochmanek and from Paul Bracken of the Yale School of Management. A number of other RAND colleagues helped our effort along the way, particularly Michael Kennedy and Frank Camm, with discussion of defense economics, and Paul Dreyer, with modifications of and assistance with the portfolio analysis tool. Michael Gilmore, Fran Lussier, and Matt Goldberg of the Congressional Budget Office were kind enough to discuss the subtleties of military compensation. Several offices in the Department of Defense provided rough estimates of actual or possible programs for the purposes of our illustrative analysis.

Acronyms and Abbreviations

The convention followed in this monograph is to avoid depending on any but the most familiar acronyms, such as DoD, OSD, and JCS. Our usual practice is to use a fuller expression but, as appropriate, to repeat the acronym parenthetically.

AFRICOM	Africa Command
AOR	area of responsibility
AT&L	Acquisition, Technology, and Logistics
BCT	brigade combat team
CBO	Congressional Budget Office
CBP	capabilities-based planning
CENTCOM	U.S. Central Command
CG(X)	guided-missile cruiser, future design
COCOM	combatant command
COIN	counterinsurgency
CONUS	continental United States
CSG	carrier strike group
DoD	Department of Defense
EBO	effects-based operations
EUCOM	U.S. European Command

FAR	flexible, adaptive, and robust
FCM	FORCES Cost Model
FYDP	Future Year Defense Plan
GPS	global positioning system
GWOT	global war on terrorism
GWS	green water squadron
HALE	high-altitude/long endurance [unmanned aerial vehicle]
HVT	high-value target
ICBM	intercontinental ballistic missile
IDA	Institute for Defense Analyses
IOC	initial operational capability
ISR	intelligence, surveillance, and reconnaissance
JCS	Joint Chiefs of Staff
JFCOM	Joint Forces Command
MALE	medium-altitude/long-endurance [unmanned aerial vehicle]
MDA	Missile Defense Agency
MEF	Marine Expeditionary Force
MTT	mobile training team
NATO	North Atlantic Treaty Organization
NDU	National Defense University
NORTHCOM	Northern Command
NPV	net present value
NSC	National Security Council
O&S	operations and support
OMB	Office of Management and Budget

OSD	Office of the Secretary of Defense
PAT	RAND's portfolio analysis tool
PA&E	Program Analysis and Evaluation
PACOM	U.S. Pacific Command
PPBE	planning, programming, budgeting, and execution
QDR	Quadrennial Defense Review Report
R&D	research and development
RDT&E	research, development, testing, and evaluation
SecDef	Secretary of Defense
SLBM	sea-launched ballistic missile
SLOC	sea line of communication
SOCOM	U.S. Special Operations Command
SOF	special operations forces
SSBN	nuclear-powered ballistic-missile submarine
SSGN	submersible, ship, guided, nuclear
STRATCOM	U.S. Strategic Command
TEAA	train, equip, advise, and assist
TRANSCOM	Transportation Command
UAV	unmanned aerial vehicle
USA	U.S. Army
USAF	U.S. Air Force
USAID	U.S. Agency for International Development
USG	U.S. government
USMC	U.S. Marine Corps
USN	U.S. Navy
WMD	weapons of mass destruction

CHAPTER ONE

Introduction

The Challenge: Resource-Informed Assessment and Recommendations

The United States is approaching a crossroad in its grand strategy and global military strategy. Since 2001, it has been involved in what has been called the global war on terrorism (GWOT) and has been engaged militarily in Iraq, Afghanistan, and elsewhere. Such engagements are quite different from what had been anticipated earlier as a steady-state posture in national military strategy and its force-sizing construct (the so-called 1-4-2-1 posture, described in Rumsfeld, 2001).[1] Also, because of the focus on ongoing conflict and stabilization efforts in the Middle East, the Department of Defense (DoD) has had to cut corners or defer efforts elsewhere. The United States will be reviewing these matters over the next two or three years and will probably then decide on a mid- and longer-term strategy. The GWOT effort, which we shall refer to as the Direct GWOT/COIN [COIN for counterinsurgency] strategy in recognition that much of what it involves is more like counter-insurgency than counterterrorism narrowly defined, may continue for many years. The form it will take and the balance between it and other DoD activities will be an issue for decision.

As strategic issues are considered and debated, the Chairman of the Joint Chiefs of Staff (JCS) will be called on to offer resource-informed

---

[1] This refers to having the capability to defend the United States itself, deter hostilities in four regions of the world, defeat two adversaries in near-simultaneous major conflicts, and defeat any one adversary decisively, which might require imposing regime change.

advice and recommendations to the Secretary of Defense (SecDef) and the President.[2] This monograph describes and illustrates a methodology intended to facilitate the framing and evaluation of resource-informed strategies and also to facilitate preparation of comprehensible summary depictions. That is, the monograph is about decision support for top-level military and civilian leaders. This might suggest that our approach would be fully "top-down" in character. In fact, our approach is more complex; it encourages a top-down flow of logic when summarizing issues, but it is built on a framework that recognizes the multifaceted nature of U.S. objectives and strategy and the central role of the military commands on which the burden for action must fall. Assessments of alternative military strategies, moreover, must reflect a level of analysis that may seem "bottom-up" to national authorities, even if it is seen as strategic and top-down by the individual commands.

So also, our approach is *not* one of unconstrained strategic thinking leading to a budget; nor is it one of budget-driven thinking. Rather, the approach encourages and enables realistic and iterative thinking about objectives, strategy, and resource implications until—at the time of final decision—the contradictions have been adequately reconciled so that the announced strategy can actually be executed within economic constraints that are considered acceptable by the American people and the U.S. Congress once they understand the consequences and tradeoffs.

Enhancing the National, Joint Perspective

Aligning Joint Responsibilities, Authorities, and Resources

One important theme of the monograph is taking a natural next step in "jointness" by characterizing and evaluating alternative national military strategies in a framework that organizes around the combatant commands (COCOMs)—i.e., around those who must actually execute strategy. This approach is intended to sharpen the process of clearly aligning responsibilities, authorities, and resources. It builds on

---

[2]   Appendix A summarizes the chairman's relevant responsibilities.

the jointness achieved since the Goldwater-Nichols act (U.S. Congress, 1986). Some aspects of the approach were also motivated by lessons learned from the private sector's large and complex enterprises, as discussed in a companion monograph.[3]

Partnership with the Military Departments

The bulk of U.S. defense planning is organized around the way in which capabilities are obtained and honed. The suppliers of the capabilities on which the nation depends for actual military operations are the military services, and the vast majority of the effort to construct and execute programs occurs within the services. That is, "Title 10 activities" dominate many practicalities. This is often construed by advocates of jointness as an unfortunate and artifactual consequence of the nation's history, especially by those who would prefer a more centralized (i.e., less service-centric) approach to planning such as can be observed in some other nations. Our own view is different. We see the military services as extraordinarily important partners in the U.S. defense enterprise. Within the U.S. services reside the deep and continuing knowledge, talents, and passions on which the nation draws constantly. The services are not just "suppliers," in the sense of being elements of a commercial marketplace that will provide what is requested and specified. The services are deeply involved in strategic planning, research, and experimentation and—far more than a decade or so in the past—their involvement is with joint contexts in mind. It is the services that look ahead, anticipate problems, suggest solutions, and ultimately develop the needed capabilities. Even when the Office of the Secretary of Defense (OSD) and the Joint Staff must override a service preference or require that some activities be accomplished jointly (as with a joint program office), it is usually to elevate the priority of activities developed within one or more services—activities that might not have been funded and encouraged without the secretary's or Joint Staff's intervention but had been conceived and subjected to experiment in the marketplace of ideas enjoyed within all of the services.

---

[3] Although this monograph and the companion piece (Gompert, Davis, Johnson, and Long, 2008) stand alone, readers may find it useful to look at them at the same time.

Within this monograph, we touch on the partnership only lightly, because our focus is the methodology for constructing and evaluating strategies and estimating their resource implications, but our companion monograph discusses the matter in more detail, drawing from the experience of large and complex business enterprises.

Need for an Integrated Portfolio Framework

Our methodology integrates an evaluation of strategies using a portfolio-management approach. The motivation is, first, that a national military strategy must be evaluated from many perspectives. Or, to put it differently, such a strategy has many components addressing different objectives. Thus, a given strategy can be conceived of as a portfolio of investments and other action items, one touching on the various objectives.[4] This is very much what portfolio management is about generally. For example, in the realm of personal finance, an individual's investment portfolio may include stocks, bonds, real estate, and money-market funds. Such variety is customary because individuals have multiple objectives such as long-term capital gain, current income, and—important—protection against the risks posed by normal financial-market fluctuations. Managing risk has been a traditional core element of portfolio work.

Our portfolio approach stems from the following principles:

1. **Integration.** We wish to be able to assess a strategy simultaneously for its likely effectiveness, risks, upside potential (not discussed in this monograph), and resource implications.
2. **Comprehensiveness.** The assessment should explicitly address each of the many high-level categories of objectives so that a proposed strategy can be assessed for its balance.

---

[4] The phrase "portfolio of" can be used as an adjective for investments, systems, or capabilities. Thus, someone building a defense program may think in terms of a portfolio of investments, but the purpose is to create a portfolio of future-commander capabilities suitable for diverse operational circumstances.

3. **Responsibleness of strategies.** Although an individual investor may choose to put all of his funds in a single stock that he is convinced will rise, the Department of Defense has the responsibility to attend to the nation's multiple military objectives and to avoid make-or-break risks. It does not have the luxury of just picking the one issue currently considered most pressing. To be sure, it can cut corners in one dimension of its activities while "plussing up" others, but it should nonetheless attend to all issues. One consequence is that *each* strategy, if it is to be "responsible," should have a concept for how and to what extent it will address multiple objectives.[5] During the Cold War, for example, U.S. military strategy focused heavily on deterring the Soviet Union, but it also included substantial capabilities for peacetime presence in many theaters of the world and for the possibility of conflicts with, for example, China, North Korea, North Vietnam, Cuba, Iraq, and other nations not controlled by the Soviet Union.

4. **Diverse resource implications.** An evaluation of a strategy's resource implications should include not only dollar costs, such as the cost of procuring a new weapon system or adding ships of the line, but also nonmonetary implications relating to, for example, the allocation of existing military resources (e.g., divisions, wings, and battle groups), the human capital (people) that constitute the military services, and the use of existing government-owned infrastructure.[6]

5. **Sound economic analysis.** The characterization of resource implications should be economically sound. In particular, it should address *all* costs—direct and indirect, immediate and deferred. Moreover, it should address all costs to the U.S. government, not just those falling under the DoD's budget. The strategies in question, after all, are *national* strategies. The Sec-

---

[5] This contrasts with posing alternatives as idealized, starkly drawn strategies that ignore considerations other than the strategies' main themes.

[6] Such issues are treated in our companion study (Gompert, Davis, Johnson, and Long, 2008).

retary of Defense and the Chairman of the Joint Chiefs of Staff serve the President in helping to develop those strategies in cooperation with other cabinet departments.

Outline of the Monograph

The remainder of the monograph is organized as follows (with the overall flow shown in Figure 1.1, to illustrate the beginning-to-end character of the methodology). For each candidate strategy, we characterize component objectives and component strategies; we then infer capability needs and draw implications for existing and future forces; next, we estimate the resource implications. Finally, we provide summary assessments of the alternative strategies in a portfolio-management framework.

Although Figure 1.1 shows where we are going as we move through the monograph, the chapter order is different because it is necessary to define some concepts and methods early, so that the reader can follow readily the subsequent flow. Thus, Chapter Two discusses our organizational approach, which we call the "operating-unit perspective." Chapter Three describes how we characterize resource implications in a straightforward and minimally burdensome way. After establishing that background, we begin the flow of Figure 1.1. In Chapter Four, we sketch a set of alternative, illustrative national strategies; we then characterize their implications with respect to combatant-command–level objectives, capability needs, and costs using the structure explained in Chapter Two. Each strategy includes a number of action steps (e.g., reallocations or purchases attempting to address the various needs). Finally, in Chapter Five we present an integrated comparison of the strategies. This comparison depends on notional and subjective evaluations of the strategies' likely effectiveness and risks and approximate estimates of their resource implications. Were the methodology applied in more depth within or for the U.S. government, the strategies would be somewhat different and probably more numerous; the evaluations could call on broad, deep analysis as well as wargaming and judgments; and the cost-estimating would draw on extensive work

Figure 1.1
Overall Flow of the Methodology

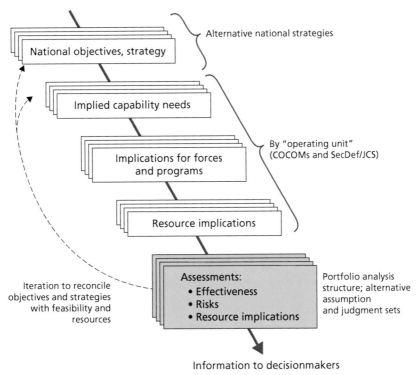

National objectives, strategy

Alternative national strategies

Implied capability needs

Implications for forces
and programs

By "operating unit"
(COCOMs and SecDef/JCS)

Resource implications

Iteration to reconcile
objectives and strategies
with feasibility and
resources

Assessments:
• Effectiveness
• Risks
• Resource implications

Portfolio analysis
structure; alternative
assumption
and judgment sets

Information to decisionmakers

RAND *MG703-1.1*

by the services and OSD's Program Analysis and Evaluation (PA&E) office.

CHAPTER TWO

The Operating-Unit Perspective

Motivations

The Logic of the Operating-Unit Framework

The foregoing section outlined a high-level approach to developing an approximate but integrated characterization of alternative defense strategies and their resource implications. The key to this methodology is viewing the combatant commands in a way analogous to how a commercial enterprise regards its operating units and, through them, plans and implements its strategy.

Broadly speaking, in a commercial enterprise, a business strategy is developed and resources are allocated to the operating units that manage the corporation's lines of business. Those units are held responsible for delivering results. The units are, literally, on the front lines of implementing the strategy, so their success or failure provides a well-tuned feedback mechanism on how well the strategy is working and whether it needs to be altered or adapted to a changing environment. In the 1970s and 1980s, businesses learned the value of organizing their strategic planning around operating units rather than production units. Doing so permitted a better alignment of allocated resources and responsibilities. It was a substantial aid in improving the productivity of the enterprises (Galbraith, 2005).

From an organizational perspective, the operating units are resourced to carry out the corporation's strategy: The operating unit has become, for a wide spectrum of enterprises, the fulcrum for linking objectives and costs.

In a somewhat like manner, the Department of Defense looks to the combatant commands to execute its strategy. These commands can be viewed as operating units to which DoD communicates its strategy and then deploys resources, primarily in the form of military forces. The forces themselves are developed and managed by the military services. In turn, COCOMs are accountable to DoD leaders for successful execution of the nation's defense strategy. By analogy to a commercial enterprise, they are DoD's vehicle for connecting resources to strategy. This is all consistent with DoD's approach to capabilities-based planning (Joint Defense Capabilities Study Team, 2004).

The Framework as a Next Step in Jointness

This approach of organizing planning around COCOMs as operating units is a natural next step in the steady move of DoD toward greater integration of forces. In the two decades since passage of the Goldwater-Nichols act, great strides have been made toward improving the degree to which military planning and operations are "national" and "joint," rather than Balkanized by military service. This has been reflected, for example, in aspects of capabilities-based planning (CBP) and effects-based operations (EBO), in the emergence of U.S. Joint Forces Command, and in substantial changes in the education of military officers as they advance through the system and through the war colleges. Today's officers are much more acquainted with "thinking joint" than were their predecessors, even though they spend most of their careers within one military service. They understand that operations will occur in joint contexts and that they must prepare accordingly. In the wake of the military crises and conflicts over the last 15 years, this is no longer an abstraction.

Properly reflecting jointness within defense planning, however, has long proved difficult. Although a succession of defense secretaries have emphasized the need to elicit and heed the requests from and views of the combatant commanders, the result has often been that the DoD receives inconsistently developed priority lists rather than full participation. One reason has been that the combatant commanders have "day jobs," attending to their daily operational responsibilities. Another reason, arguably, has been that when commanders' inputs have been requested, it has been largely a matter of eliciting command-specific (even command-parochial) suggestions rather than including the commanders effectively in an overall dialogue about strategy.

Given this background of progress toward jointness, tainted by a lack of success in successfully integrating COCOMs into the planning process by requesting their inputs, we see the concept of organizing the portfolio-management structure of strategic planning around COCOMs as something with considerable potential for moving jointness forward. The process would compel DoD planners to be deliberate about selecting objectives for each COCOM. The aggregation of these objectives should, in a gross sense, reflect a strategic plan for U.S. defense.

The objectives should in turn inform planners of the capabilities and thus of the forces to be provided to the COCOMs. As joint engagement around the globe in peacetime becomes increasingly critical to a global strategy, the systematic examination of COCOM objectives and the requisite resources provides discipline to the planning process.

Conceptually, this is not entirely new. Achieving the goals of a global defense strategy depends even today on whether goals can be faithfully translated into objectives and whether the planning process provides the COCOMs with the wherewithal to achieve them. The innovation is that the accounting for such matters explicitly within the planning, programming, budgeting, and execution (PPBE) system would systematically focus on the organizations that are responsible for implementing the strategy. Further, the approach might increase the speed and consistency with which COCOM needs are addressed. And, over time, it might increase the DoD's overall effectiveness and efficiency because the consistency or inconsistency of resource decisions with strategic intent would be clearer.

Cautions

The analogy between business-world operating units and COCOMs is far from perfect. The COCOMs do not control sizable operating budgets as do the operating units in a commercial enterprise. In fact, they control less than 1 percent of the DoD budget. Control of the bulk of the defense budget—approximately 85 percent—rests with the services (defense agencies accounting for most of the remainder). Thus, the COCOMs find it more difficult to respond quickly to changes in the environment than do operating units in commercial enterprises. Moreover, the PPBE system is notoriously slow and cumbersome and badly configured for responding to rapid changes in the security environment. Most important, perhaps, the success of a national strategy depends not only on the military-strategy component but also on components managed by the Department of State and other cabinet departments. Indeed, an individual combatant commander may not be able to achieve his military objectives without effective parallel activities by those other departments.[1]

Current COCOMs and "National Command"

Current COCOMs

Our study used the ten current COCOMs as the focus of the analysis (Table 2.1). Of the ten commands, six are defined by their regional or geographic responsibilities (see Figure 2.1); the others have global or functional responsibilities. The regional COCOMs are well suited to serve as vehicles for the analysis of alternative global strategies, whereas some functional COCOMs—Transportation Command (TRANSCOM) and Joint Forces Command (JFCOM) in particular—are best treated as "capability providers" that enable the regional COCOMs to meet their peacetime and combat objectives. SOCOM (U.S. Special Operations Command) is an exceptional case. It acts as a capability provider and also has an independent global strategic role, most notably as DoD's lead command for GWOT. Strategic Command (STRATCOM) can also blur the line between capability provider and indepen-

---

[1] To put the matter differently, the logical "operating units" would need to be organizational entities that do not exist in the U.S. government and that would report to the National Security Council (NSC). The complications introduced in national security and other aspects of government by organizational strains is considerable, and—some would say—worsening because of global trends (Bracken, 2007).

Table 2.1
The Combatant Commands

Command	Abbreviation	Area of Responsibility
Regional Combatant Commands		
U.S. Central Command	CENTCOM	Middle East (Egypt through the Persian Gulf into Central Asia, Pakistan)
U.S. Pacific Command	PACOM	Asia-Pacific region, including Hawaii, Alaska, and India
U.S. European Command	EUCOM	Europe, Russia, and Israel and surrounding waters
U.S. Africa Command	AFRICOM	Africa, excluding Egypt
U.S. Southern Command	SOUTHCOM	Central and South America and surrounding waters
U.S. Northern Command	NORTHCOM	Continental United States (CONUS)
Functional Combatant Commands		
U.S. Strategic Command	STRATCOM	Worldwide missions
U.S. Special Operations Command	SOCOM	Worldwide missions
U.S. Transportation Command	TRANSCOM	Provider of global mobility by air and sea
U.S. Joint Forces Command	JFCOM	Joint provider of forces and training

dent actor. We treat both SOCOM and STRATCOM as operating units in the ensuing analysis, and we treat TRANSCOM and JFCOM as included in National Command (described in the next section).

Portions of the DoD budget (research and development, joint logistics, and a number of others) will always be managed centrally, as in most commercial enterprises, and not carried out with a focus on particular COCOMs.[2] These are treated as resources that rise or

---

[2] In this monograph, "central" refers to national-level activities (often global in scope), such as occur in the Joint Staff and OSD, or in analogous organizations of the military departments (e.g., offices of the service chiefs or secretaries). The service chiefs and secretaries are

Figure 2.1
Geographic Responsibilities

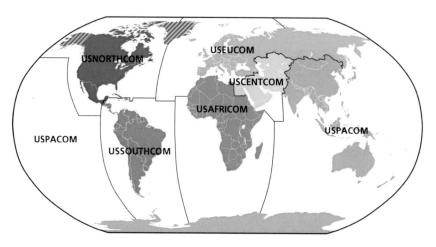

SOURCE: DoD News Briefing, April 17, 2002.
NOTES: This map shows AFRICOM's objective AOR in green. The command has
not yet assumed responsibility for all of the demarcated area. Alaska Command
is a sub-unified command under PACOM.

RAND *MG703-2.1*

fall depending on the overall thrust of the strategy, not on particular
demands from the COCOMs.

A National Command for Analytic Purposes

Although no COCOM exists for the national- or global-command
function, that function is an essential aspect of the overall system,
even if notional. Such a command could provide support of all kinds
to the regional COCOMs, from training and transportation to the
forces necessary to conduct emergent operations. These forces and support come from the functional combatant commands and from those
portions of the force structure that are not assigned to any particular
regional COCOM. In practice, all of the military force structure could

---

"top level" or "central" in our context, whereas service components, such as U.S. Air Forces
Europe or the Pacific Fleet, are analogous within service chains to COCOMs. In another
context, "central" would refer to DoD headquarters, rather than services. That is not our
usage.

be shifted (at varying speeds and expense) from COCOM to COCOM to respond to pressing needs. Under law, the National Command function is best associated with the Secretary of Defense. The defense secretary, in turn, depends on the Chairman of the Joint Chiefs of Staff, who relies on the Joint Staff (e.g., its directorate for operations, J-3). The Joint Staff draws on the regional commands and the functional commands (including JFCOM and TRANSCOM). The National Command function includes global planning (e.g., for simultaneous crises). For the purposes of methodology, this National Command is also understood to maintain a pool of reserves that can be used wherever needed, including to reinforce forces of a regional COCOM.

With this background, let us now identify and characterize some strategies and begin applying the methodology sketched in Chapter One.

CHAPTER THREE

Characterizing Alternative Strategies in Terms of Implications for Operating Units (COCOMs)

In this chapter, we illustrate how our approach to estimating the resource implications of alternative strategies could work. A critical element of the discussions would be an estimate of the resource implications of any candidate strategy. In the following chapter, we posit strategies that typify those that might be developed by OSD or the Joint Staff for consideration by the Secretary of Defense and Chairman of the Joint Chiefs of Staff.

Overview

Expressing Strategy and Goals

The resource implications of an alternative strategy are derived from a logical flow that begins with a clear statement of the strategy. This is preceded by a short statement of the premise, or the world-view that motivates the strategy.

- What conditions are emerging that pose challenges to our national security?
- Why are they important and why do they need to be addressed as a priority by our national security strategy?

The *strategy statement* indicates how these challenges will be addressed. It is focused enough to provide force planners with a clear vector for making force-structure and programmatic choices but inclusive enough to address the broad spectrum of enduring U.S. national security imperatives.

Next, the *goals* of the strategy are articulated in output terms. What end state does the United States seek in choosing and embarking on a particular strategy? The final task in developing a strategy is to examine the goals and develop the best *approach* for achieving them.

Planning for Adaptiveness

Developing a strategy in pursuit of goals and objectives is not as straightforward as one might think from textbook descriptions—in large part because only some aspects of the future are controllable. The United States can influence, but not control, the behavior of other nations and organizations (including nonstate actors such as al Qaeda). Moreover, "things happen." As discussed in a long string of RAND studies over the last 15 years, as well as in the academic and business literature, a core element of strategy needs to be *planning for adaptiveness*.[1] It is useful to distinguish between two kinds of adaptiveness:

Strategic adaptiveness: the ability to adjust effectively to changes that may occur in the international and domestic environments, whether those be such changes as the emergence of new competitors or adversaries, the resolution of long-standing problems, large shifts in the need for or availability of resources, or the emergence of new technologies. Strategic adaptiveness typically refers to changes over a period of years.

Operational adaptiveness: the ability to adjust operations or operational concepts quickly to deal effectively with variations or changes in adversary strategy or tactics, the presence or absence of allies, or other events. Operational adaptiveness applies, for example, during a crisis or war.

Some principles for planning under uncertainty by encouraging adaptiveness include explicitly preparing in some detail for foreseeable

---

[1]   RAND strategic thinking on these matters has emerged over the last 15 years (Davis, 1994b; Davis, Gompert, and Kugler, 1996; Davis, 2002; Johnson, Libicki, and Treverton, 2003; Dewar, 2003). A recent book puts RAND thinking into the broader context of organizational effectiveness generally (Light, 2004), contrasting it with the business literature.

contingencies and maintaining more general hedges, such as military forces in reserve, slack in command and control systems, redundancy, and multipurpose systems and units. For the purposes of this study, this type of thinking translates into recognition that, for each strategy considered, we should identify the assumptions on which it is based that are somewhat fragile, as well as the capabilities that would be needed in the event the assumptions fail. More tangibly, each strategy should explicit include concepts for adaptation and identify "requirements" for resources that would facilitate such adaptation. Let us now return to the basic flow.

The Analytic Flow

The analytic flow (Figure 2.1) is as follows: strategy statement to goals, to approach, to objectives of the COCOMs, to capabilities required by the COCOMs, to choices of forces and other programs, and, finally, to costing (and other characterizations of resource implications). This

Figure 3.1
From Strategies to Resource Implications

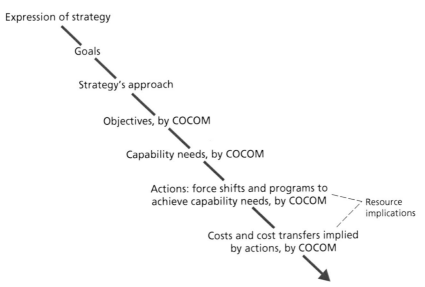

approach is illustrated below for an analytic baseline strategy and for three alternative strategies. Each is plausible and has been discussed by responsible participants in the national-security community. The strategies, then, are titled

1. Analytic Baseline
2. Direct Global War on Terrorism/Counterinsurgency (Direct GWOT/COIN)
3. Build Local, Defend Global
4. Respond to Rising China.

The Analytic Baseline provides a vehicle for expressing enduring strategic aims and, later, core resource allocations. Strategies 2–4 are our first-cut versions of concepts that are possible alternatives; they provide a broad enough range of strategic focus to illustrate the methodology and raise interesting, relevant issues of the day. With this overview of the flow, let us discuss what we mean by *objectives, capability needs, actions,* and *costs.*

Elements of the Process

Operating-Unit Objectives

Given an articulated overall defense strategy, the Department of Defense would look to the combatant commands (COCOMs) to implement many elements of it, in rough analogy to the way a commercial enterprise would carry forward its strategy. This phase in the analysis demands that objectives be set for each operating unit which, if achieved, would lead to successful implementation of the strategy. This step, of necessity, precedes the determination of capabilities and therefore resources (military forces and security assistance funding for the most part) to be provided to the COCOMs.[2]

---

[2] COCOMs do not literally control security-assistance funding but they are typically a very important part of the effort to define and implement security assistance.

Capabilities Needed to Meet Operating-Unit Objectives

With the objectives of each COCOM established, DoD force planners would then determine what capabilities are implied beyond those called for in the baseline force structure and the Future Year Defense Plan (FYDP). The objectives are analyzed COCOM by COCOM and the needed capability is stated clearly in output terms.[3] This step, as with the determination of objectives, precedes the calculation of what resources are to be provided to the COCOMs. For some strategies, certain COCOMs will need an increase in capabilities whereas others might need fewer than provided for in the baseline.

Operating-unit objectives are drawn from strategies whose successful implementation can include diplomatic and economic initiatives as well as military capabilities. The strategy statement (described above) captures important characteristics of a given strategy and attendant objectives (e.g., does the United States seek to contain or engage a particular country?). The military capabilities and the supporting programs and force shifts that flow from the strategy would typically have to be supplemented by other instruments of U.S. national power.

Programs and Force Shifts to Develop Needed Capabilities

With an assessment of the capabilities needed by the several COCOMs complete, force planners can identify the programs and force shifts that they believe will effectively, and cost-effectively, deliver those capabilities. Some COCOMs will require programs—additional ground units, aircraft, security-assistance spending, and the like—beyond those in the baseline to provide the extra capabilities implied; some, under certain strategies, could require less and the programs would be backed out of the force structure/FYDP.[4] Capabilities can also be added or subtracted by shifting forces from one COCOM to another. The meth-

---

[3] Establishing the nature of these capabilities is by no means trivial. Several levels of analysis will be needed to establish the type and scale of capabilities necessary to meet a given objective. Reasonable people can and will disagree on what constitute sufficient resources to meet the tasks at hand.

[4] When the purpose of analysis includes making tradeoffs under economic constraints, the baseline should be adjusted to reinstate aspects of the program that need to be reconsidered.

odology makes explicit this addition, subtraction, or redistribution of forces and other programs among the COCOMs.

Resource Implications

Implications for Force Posture and Programs. In this step of the process, costing experts cost out the programs. This life-cycle costing[5] is done COCOM by COCOM for programs outside the force structure or FYDP. The programs are phased in a manner consistent with DoD's ability to launch the programs. For example, an expansion in naval forces cannot be effected immediately. A decade or more could pass before a measurable expansion in the number of capital ships in the fleet were accomplished.

The costs are built up from the component parts: personnel, operations and support (O&S), procurement, and research and development (R&D). They are attributed to the service (or in a few cases, the agency) that has the responsibility to execute the program. The results are therefore transparent at the level of DoD strategic planning. They provide an audit trail that indicates

- increases, decreases, or shifts of resources among the COCOMs
- increases, decreases, or shifts of resources among the military services
- challenges in meeting the resource requirements (e.g., impending large expenses, which are sometimes called gathering "bow waves")
- changes in resource requirements in DoD and other cabinet departments such as the Department of State
- a 20-year summary estimate (or other measures as mentioned above) of the increase, or decrease, in the cost of an alternative strategy relative to baseline expenditures.

---

[5] The cost estimates discussed below do not include disposal costs of weapons, toxic wastes, and the like.

The next chapters describe in some detail the execution of the methodology for the three strategies cited as alternatives to the Analytic Baseline.

Extraordinary Costs of Operations. The costs of defense are high even in ordinary times: Maintaining force structure, infrastructure, and forward deployments is expensive. In times of war or other unusual periods (such as today), however, "extraordinary costs" (costs beyond those of the core defense program) come into play. Traditionally, U.S. defense planning has not dealt with these—adopting the view that such extraordinary costs would be dealt with as necessary when the time comes—independent of what strategies had been pursued previously. The United States did not prepare fiscally years in advance for the Korean or Vietnam wars; nor did it include in the budgets of the late 1980s the anticipated expense of a war after Saddam Hussein invaded Iraq. And, more recently, the unusual expense of actually conducting the wars in Afghanistan and Iraq, and the counterinsurgency efforts that continue to this day, have been covered by supplemental appropriations. These include funds for deployment, combat pay, special new equipment, and overhaul of heavy equipment as it comes back from operations.

It can be argued that such extraordinary costs cannot reasonably be built into the assessed costs of a strategy in a strategic-planning exercise focused on the mid to long term. However, that is a reasonable argument only to the extent that reasonable expectations on such matters would be invariant across the strategies considered. That is not the case for the strategies considered in Chapters Four and Five. Thus, we shall address extraordinary costs explicitly.

Other Kinds of Costs. Although we do not address them in any depth in this monograph, strategic planning must also consider a number of resource implications or nonmonetary costs when comparing strategies. Sometimes, these correspond to constraints, such as the inability of the U.S. industrial base to build ships rapidly without an expansion possible only in times of emergency, such as World War II, or the inability to recruit and train as many ground-force personnel as might be desired without lowering standards or reinstituting the draft. Constraints can typically be eased on the margin with economic incentives (e.g., enlistment and retention bonuses), but there are limits. In some cases, they can be eliminated over time, as when the United States invested in the industrial infrastructure to build precision weapons. Sometimes the constraints can be eased or eliminated by drawing on the infrastructure of other countries. However, that may require creating undesirable dependences or a web of politically and strategically complicated relationships. Our companion monograph discusses these matters more extensively (Gompert, Davis, Johnson, and Long, 2008).

CHAPTER FOUR

Application to Some Illustrative Strategies

This chapter works through the approach, step by step, beginning with the statement of some alternative strategies that, although illustrative, relate well to current-day issues of grand strategy. Discussion is deliberately abbreviated, almost telegraphic, since the focus of this monograph is the methodology to determine the resource implications of different strategies, not developing the strategies themselves in detail.

Characterizing the Strategies

Premises of the Alternative Strategies

Direct GWOT/COIN. Extremist Islamist insurgency is a worrisome and growing phenomenon that threatens the homeland and important U.S. interests in sensitive regions such as

1. energy-producing countries in the Gulf and North Africa
2. countries straddling key lines of communication such as Indonesia
3. important Muslim-majority allies and partners such as Pakistan and Turkey.

Islamist insurgencies also threaten Israel, to which the United States has long-standing security responsibilities.

This is a relatively new phenomenon, at least in its intensity. Although the United States and its allies have well-honed approaches to dealing with symmetric competitors, they are still feeling their way on how to cope with this largely new challenge. It seems unlikely that local states alone will be able to deal with the threat for reasons that include (depending on the state) poverty, low levels of democratization and perceived legitimacy, incompetence, and corruption. As a result, *the United States should plan on continuing, direct intervention to assist in GWOT/COIN operations.*

Build Local, Defend Global. Although the concerns of the GWOT/COIN strategy are valid, eliminating insurgent threats through large-scale U.S. military operations has proven to require a very large investment in U.S. forces and to be extremely expensive.[1] All this, with no guarantee of success. Indeed, the presence of U.S. military forces conducting operations on the territory of other states can create a strong backlash, if adversaries depict such operations as unwanted "occupation" of their lands and as a U.S.-led "war against Islam." Hence, for a variety of reasons, local instability is best dealt with by local capacity, which implies a strategy of investing heavily to build and sustain those local capacities.[2]

Respond to Rising China. Although the Islamist threat is undeniable, the preeminent challenge is the rise of China. Chinese diplomatic, economic, technological, and military power cannot help but alter the strategic landscape. Unless the United States takes proactive measures, this expansion could take place at the expense of U.S. interests. A strong U.S. stance, in the Pacific and globally, will lay the foundation for a stable, peaceful (albeit competitive), long-term relationship with China. The Islamist challenge, although substantial, can probably be dealt with by supporting the efforts of local countries and without greater investments than are already part of the Analytical Baseline. Further investment would probably not pay off sufficiently to make it worthwhile.

---

[1]   Current estimates are that operations in Iraq and Afghanistan over the period 2001–2017 will cost about two trillion dollars (Congressional Budget Office, 2007b).

[2]   See Grissom and Ochmanek (2008) and Gompert, Gordon, Grissom, Frelinger, Jones, Libicki, O'Connell, Stearns, and Hunter (2008) for further development of the imperative that the United States serve as an enabler of COIN operations in foreign countries, rather than as the principal actor.

Contrasting Goals, Approaches, and Preparations for Adaptation

Against a background of conflicting premises, the strategies have contrasting goals and approaches, as described in Table 4.1.

The Possibility of Failure and the Need for Strategic Adaptiveness

Any of the strategies could fail. The Direct GWOT/COIN strategy might prove extremely costly and demanding of ground forces. Alternatively, the Islamist threat might ease but a Chinese build-up in East Asia and the Western Pacific might develop more rapidly than expected. The Build Local, Defend Global strategy might fail because the governments of the local partners fail to gain sufficient legitimacy or prove incompetent. The United States would not be well prepared for more manpower-intensive intervention. And, as with the first strategy, problems might arise rapidly in PACOM. Focusing on responding to a rising China might fail because of insufficient attention to the Islamist problem, which might at some point explode. Such a focus might also trigger an unintended arms race with China, which would be very costly. Or, the strategy might prove inadequate to deal with a rapid build-up of modern Chinese forces threatening Taiwan and more general U.S. interests in the region. In all of these cases (and others too numerous to enumerate), then, strategy and programs would have to change. Fortunately, the Analytic Baseline strategy itself provides for substantial future capabilities, including considerable modernization. It also includes extensive R&D that, we hope, will lay the basis for strategic adaptations that might be necessary.

Objectives of COCOMs

The key to success of the Direct GWOT/COIN strategy is the capable performance of CENTCOM, although other COCOMs (and other agencies of government) have key contributions to make. The Build Local, Defend Global strategy, although also focused on the Islamist threat, seeks to improve the capabilities of partner countries worldwide. The Respond to Rising China strategy, of course, is largely focused on PACOM. Table 4.2 provides some of the more salient objectives of the

Table 4.1
Characterizing the Alternative Strategies

Strategy	Goal	Approach to Core Goal	Hedges
Direct GWOT/ COIN	Diminished threat from, capability of, and support for Islamist insurgents acting against U.S. interests.	Improve COIN operations in the Muslim world with an approach that focuses on U.S. military operations tailored to the manpower-intensive task of countering an active insurgency. Other facets include special operations against high-value targets (HVTs) improved capability of indigenous security forces. Need for military interventions in Islamic world expected.	Hedge against the possibility of other global crises, particularly confrontation with China. Maintain the Analytic Baseline strategy's plans for a strong U.S. presence in East Asia, primarily with naval and air forces, and continue with robust R&D efforts suitable for deterring or dissuading peer or near-peer competitors.
Build Local, Defend Global	Indigenous forces able to counter nonstate threats, multilateral frameworks for regional problems, and U.S. forces freed to focus on global commons and other global interests.	Help indigenous security forces in partner countries develop competence to handle nonstate threats. Foster multilateral cooperation with capable allies and partners. Plan on much-expanded security and foreign assistance. Intervene directly only as necessary to prevent a strategic shift in a vital region defeat threats to the free flow of goods or access to energy sources protect an ally.	Hedge against the possibility of a direct threat to Gulf energy supplies by maintaining forces and presence adequate for intervention. More globally, hedge against an acceleration of challenges in East Asia and elsewhere by maintaining the Analytic Baseline strategy's plans for a strong U.S. presence in East Asia, primarily with naval and air forces, and with robust R&D efforts suitable for deterring or dissuading peer or near-peer competitors.
Respond to Rising China	Responsible China constructively engaged in international affairs, deterred from acts of military intimidation or coercion.	Avoid vacuums and deter and dissuade in ways appropriate to benign competition between powers that need not become adversaries. Plan military capabilities to ensure that the United States could prevail in any plausible conflict in the Western Pacific or globally. Encourage Sino-American cooperation in areas of common interest such as counterterrorism and sea line of communication (SLOC) security.	Maintain the capability to intervene to protect vital interests in the Middle East and elsewhere.

Table 4.2
Notable COCOM Objectives for the Analytic Baseline Strategy

Command	Core Objectives
CENTCOM	Promote stability of allied governments and prevent creation of terrorist havens Assure access to energy sources Contain and keep pressure on Iran
PACOM	Deter North Korean attack Deter Chinese attack on Taiwan Protect SLOCs Build up indigenous COIN capacity Prevent growth of terrorist capabilities
EUCOM	Maintain a strong security partnership with European allies Enhance European focus on and capabilities for global counterterrorism and COIN operations
AFRICOM	Improve indigenous security forces and promote good governance Conduct limited direct operations against terrorists
SOUTHCOM	Promote stability of local governments and prevent creation of terrorist havens Improve regional allies' capabilities to conduct counterinsurgency and counterdrug operations
NORTHCOM	Prevent terrorism against U.S. territories Support civil authorities in counterterrorism and disaster response
STRATCOM	Provide assured global nuclear deterrent Provide national and theater missile defense Provide national aerospace and cyberspace security
SOCOM	Support national-level objectives where highly focused or covert action is needed
National Command	Deter support for terrorism Deter the rise of potential military competitors Provide rapid global support to other COCOMs and to allies Limit the proliferation of weapons of mass destruction (WMD) and other advanced weapons technologies

operating-unit objectives for the Analytic Baseline strategy—omitting many others for brevity (each COCOM has many enduring objectives that are common to all the strategies). Tables 4.3–4.5 highlight objectives particularly relevant to the alternative strategies. *Note that the objectives of the baseline strategy are assumed to apply to the alternative strategy unless otherwise noted.* As an example, under all strategies, the United States must be able to intervene to protect oil supplies or to

Table 4.3
COCOM Objectives for the Direct GWOT/COIN Strategy

Command	Highlighted Objectives
CENTCOM	Execute direct counterinsurgency operations Identify and strike terrorist targets Build up indigenous COIN capacity
PACOM	Identify and strike terrorist targets Build up indigenous COIN capacity
EUCOM	*No change from baseline*
AFRICOM	Identify and strike terrorist targets Build up indigenous COIN capacity
SOUTHCOM	*No change from baseline*
NORTHCOM	*No change from baseline*
STRATCOM	*No change from baseline*
SOCOM	*No change from baseline*
National Command	*No change from baseline*

deal with a North Korean attack of South Korea or attempted Chinese coercion of Taiwan.

Capabilities Needed by Operating Units

When decisionmakers reflect on changing a course of action or, in this case, adopting a new defense strategy, they are typically interested in the change from the existing baseline, asking

> *"If we move forward with a more aggressive defense strategy, one that does not depend on strong contribution from allies, how much more will that cost?"*

or

> *"If we pull back and let local forces take responsibility for their own security, how much could we save?"*

Table 4.4
COCOM Objectives for the Build Local, Defend Global Strategy

Command	Highlighted Objectives
CENTCOM	Build the capability of indigenous security forces, both on shore and at sea Provide direct military support for local allies as needed Enable actionable intelligence for limited strikes on terrorist targets and early warning of incipient crises Improve the capacities of local governments and economies
PACOM	Build the capability of indigenous security forces, both on shore and at sea Enable actionable intelligence for limited strikes on terrorist targets and early warning of incipient crises Improve the capacities of local governments and economies
EUCOM	Improved the capability to rapidly deploy forces out of area Increase the contribution of allies to security in the Mediterranean and Atlantic
AFRICOM	Build the capability of indigenous security forces, both on shore and at sea Improve the capacities of local governments and economies
SOUTHCOM	Build the capability of indigenous security forces, both on shore and at sea Improve the capacities of local governments and economies
NORTHCOM	*No change from baseline*
STRATCOM	Partially compensate for the drawdown of forward-deployed forces
SOCOM	Enhance support to COCOMs in the training of indigenous forces and direct action
National Command	Maintain the capacity to surge forces forward in those cases where enhanced direct-action capability is needed

In what follows, we take the illustrative strategies and estimate needed capabilities, programs, and costs as an increment (or decrement) from the baseline described above. Several levels of analysis are critical to getting this part of the analysis right—operating-unit objectives may in some cases indicate a clearly needed set of capabilities, but typically more than one set of capabilities can address a given objective. This is not new. Defense planners routinely wrestle with precisely this analytic challenge. This approach will not resolve that challenge. However, it does provide a useful analytic framework: clearly stating

Table 4.5
COCOM Objectives for the Respond to Rising China Strategy

Command	Highlighted Objectives
CENTCOM	Build the capability of indigenous security forces and governments Enhance SLOC security
PACOM	Maintain the capacity to establish sea control in the Western Pacific Enhance SLOC security Strengthen local alliances and partnerships and engage with China on issues of common interest Build the capability of indigenous security forces and governments
EUCOM	Increase the contribution of allies to security in the Mediterranean and Atlantic
AFRICOM	Promote political and economic progress independent of China Build the maritime security capabilities of local allies and partners
SOUTHCOM	Build the maritime security capabilities of local allies and partners
NORTHCOM	*No change from baseline*
STRATCOM	Enhance national and theater missile defense Enhance global intelligence, surveillance, and reconnaissance (ISR) capabilities Prepare to compensate for potential loss of space-based ISR and communications assets
SOCOM	Selectively employ direct action for limited periods in pursuit of HVTs
National Command	Provide ground, air, and maritime surge capability to support potential high-intensity conflict with near-peer competitor

capabilities that match the operating-unit objectives that in turn provide a foundation for identifying programs and force shifts to provide these capabilities. The resource implications of the strategy can then be expressed through these programs and force shifts in a way that links the costs and savings back to capabilities and so to objectives.

Following the flow of methodology discussed above, the next step is to characterize the capabilities needed by the COCOMs to achieve their objectives. Tables 4.6–4.8 describe these succinctly. The capabilities listed in these tables are not wholly comprehensive. Rather, they emphasize differences from the baseline (as implied in Table 4.1 and Table 4.9) particular to the strategy in question; baseline capabilities are assumed.

Table 4.6
COCOM Capabilities Needed for the Direct GWOT/COIN Strategy

Command	Highlighted Capabilities Needed
CENTCOM	Ground combat forces to sustain lengthy COIN campaigns Improved capability to train, advise, and develop local forces Improved strike capability against HVTs
PACOM	Improved capability to train, advise, and develop local forces Improved strike capability against HVTs
EUCOM	*No additional capabilities needed*
AFRICOM	Improved capability to train, advise, and develop local forces Improved strike capability against HVTs
SOUTHCOM	*No additional capabilities needed*
NORTHCOM	*No additional capabilities needed*
STRATCOM	*No additional capabilities needed*
SOCOM	Enhanced training, advisory, and direct-action capabilities in CENTCOM, PACOM, and AFRICOM
National Command	*No additional capabilities needed*

Actions: Programs and Force Shifts to Address Capability Needs

An Analytic Baseline

Establishing a baseline of forces provides a framework in which increments and decrements can be made to provide the capabilities required to underwrite a change in strategy. To illustrate the technique, the major components of active duty forces are wholly "allocated" to the several COCOMs in this section, as shown in Table 4.9 and Figure 4.1. This is a two-step process.

First, the allocation is made based on the current commitments of the armed forces, which in turn reflect the strategic choices made since the turn of the century. These choices were reflected also in Table 4.1, which gave the baseline strategic objectives. Next, the allocation is amended to project it out through the entire 2009–2028 period.

Table 4.7
COCOM Capabilities Needed for the Build Local, Defend Global Strategy

Command	Highlighted Capabilities Needed
CENTCOM	Substantial increase in training and advisory teams to build local COIN capacity Sufficient ground combat forces to deter regional competitors and to support allies if indigenous forces are overwhelmed Enhanced naval presence to improve SLOC security and partner with local forces Improved ISR Substantial financial assistance to support local security forces, government capacity-building, and economic development
PACOM	Additional training and advisory teams to build local COIN capacity Enhanced naval presence to improve SLOC security and partner with local forces Improved ISR
EUCOM	Improved capability to rapidly deploy forces out of area
AFRICOM	Substantial increase in special operations forces (SOF) training and advisory teams to build local COIN capacity Improved SOF direct-action capability Substantial financial assistance to support local security forces, government capacity-building, and economic development
SOUTHCOM	Increase in SOF training and advisory teams to build local COIN capacity Improved SOF direct-action capability Substantial financial assistance to support local security forces, government capacity-building, and economic development
NORTHCOM	*No additional capabilities needed*
STRATCOM	Improved long-range strike capability
SOCOM	Enhanced ability to respond to emerging direct-action needs
National Command	Training and advisory teams available to bolster efforts of regional commands as needed Sufficient ground forces to reinforce forward-deployed units in event of crisis Substantial ISR surge capacity

Not surprisingly, in recent years, the great bulk of ground forces has been oriented toward CENTCOM. Air and naval forces are more globally distributed, although they too have been more heavily engaged in CENTCOM operations than was the case before the invasions of Afghanistan and Iraq. The 2009–2028 baseline assumes a sizable draw-

Table 4.8
COCOM Capabilities Needed for the Respond to Rising China Strategy

Command	Highlighted Capabilities Needed
CENTCOM	Enhanced naval presence to improve SLOC security and partner with local forces Sufficient ground combat forces to deter regional competitors and to support allies if indigenous forces are overwhelmed Financial assistance to regional partners to build partner capacity to conduct COIN
PACOM	Substantially enhanced naval presence to establish sea control Enhanced capability in littoral warfare Enhanced capability to partner with local maritime forces Increased medium-range strike capability Increased stealthy strike capability Financial assistance to regional partners to build government capacity, promote economic development, and contend with Chinese influence
EUCOM	*No additional capabilities needed*
AFRICOM	Financial assistance to regional partners to build government capacity, promote economic development, and contend with Chinese influence
SOUTHCOM	Enhanced capability to partner with local maritime forces
NORTHCOM	*No additional capabilities needed*
STRATCOM	National missile defense capable of dealing with limited attack Enhanced repositionable ballistic-missile defense capability Improved long-range stealthy strike capability
SOCOM	*No additional capabilities needed*
National Command	Sufficient ground forces to reinforce forward-deployed units in event of crisis

down in U.S. forces in Iraq and Afghanistan but recognizes that unless there is a significant change in strategy, those countries, and the region in general, will continue to be a top priority for U.S. forces. Continued, but diminished, responsibilities in Iraq and Afghanistan, in addition to the need to respond quickly to other emerging regional threats, result in an enduring heavy orientation of forces to CENTCOM.

Over the same period, the trend of reorienting forces from EUCOM toward other COCOMs that began with the end of the Cold War has continued. This is not to say that EUCOM's role is

Table 4.9
Force Structure of the Analytical Baseline

Program	Command	Service
33 (18/15) brigade combat teams (BCTs)	CENTCOM	U.S. Army (USA)
33K (10/23) SOF troops	CENTCOM	Multi
2 (0/2) Marine Expedition Forces (MEFs)	CENTCOM	U.S. Marine Corps (USMC)
6 (1/5) carrier strike groups (CSGs)	CENTCOM	U.S. Navy (USN)
13 (5/8) combat wings	CENTCOM	U.S. Air Force (USAF)
$163B foreign and security assistance over 20 years	CENTCOM	Other U.S. government (USG)
6 (3/3) BCTs	PACOM	USA
6K (3/3) SOF troops	PACOM	Multi
1 MEF	PACOM	USMC
3 (2/1) CSGs	PACOM	USN
7 (5/2) combat wings	PACOM	USAF
$7B foreign and security assistance over 20 years	PACOM	Other USG
4 (1/3) BCTs	EUCOM	USA
2 (1/1) CSGs	EUCOM	USN
3 (1/2) combat wings	EUCOM	USAF
$20B foreign and security assistance over 20 years	EUCOM	Other USG
$50B foreign and security assistance over 20 years	AFRICOM	Other USG
$33B foreign and security assistance over 20 years	SOUTHCOM	Other USG
980 nuclear missiles (intercontinental ballistic missiles [ICBMs] and sea-launched ballistic missiles [SLBMs])	STRATCOM	USAF/USN
National Missile Defense Program	STRATCOM	Multi
21 BCTs	National Command	USA
26K SOF troops	National Command	Multi
2 MEFs	National Command	USMC
7 CSGs	National Command	USN
12 combat wings	National Command	USAF

unimportant. It plays a critical role in maintaining a robust engagement with U.S. North Atlantic Treaty Organization (NATO) allies, which remain the most important grouping of nations that largely share common security goals with the United States. Still, for the 2009–2028 period, EUCOM's area of responsibility is projected to be relatively stable. Cross-border aggression is unlikely, so the baseline does not project a heavy requirement for standing forces.

The forces deployed in and oriented toward PACOM have been relatively stable for about four decades. Only recently have the heavy requirements for forces in CENTCOM resulted in a refocusing of forces traditionally oriented to PACOM to CENTCOM. The baseline projects that current approximate force levels in PACOM will be relatively stable.

AFRICOM, SOUTHCOM, and NORTHCOM have not had a great requirement for forces from the active component, and with the increased requirement in CENTCOM, resources available to them have been even more limited. Their baseline requirement for resources is not projected to change significantly. Consequently, these COCOMs are not assigned major active component forces in the baseline, although some lesser resources would be devoted to them.

STRATCOM has responsibility for operation of U.S. strategic nuclear forces, for missile defense, and for important aspects of networking. Its inventory of offensive delivery systems and warheads has been steadily shrinking in accordance with the START treaty regimes, and the inventory is projected to continue to decrease gradually.

There were no strategy-driven increments or decrements of offensive nuclear forces in the analysis. However, there is growing interest in (and differences of opinion about) ballistic-missile defense. The baseline includes $9 billion a year for missile defense, roughly the projected annual budget of the Missile Defense Agency. This money will pay for progressively more capable multitiered defense against limited threats to U.S. allies, U.S. forces deployed abroad, and the U.S. homeland.[3]

---

[3] The baseline would also include any additional expenses to operate ballistic-missile defenses above and beyond what can be covered in the budget of the Missile Defense Agency.

In all strategies, some forces were held in strategic reserve to be allocated to the COCOM that had the most demanding requirement or simply to respond to surprises. These forces are held in a National Command until they have to be deployed to a COCOM. In the meantime, these forces are indicated as "earmarked" for the COCOM where they are most likely to be employed.

The baseline distribution of forces is shown in Table 4.9 and Figure 4.1.[4] In some instances in this table and figure, the number of units is followed by an allocation, in parentheses, of two numbers separated by "/". The number before the "/" indicates the forces that are specifically oriented to the COCOM and the number after it is the number of units held by National Command earmarked to that COCOM. The units listed after the "/", then, are listed again under National Command.

CENTCOM. The operations in Iraq and, to a lesser extent, Afghanistan have generated a large requirement for all types of forces, but ground forces in particular. This baseline assumes a sizable drawdown in U.S. forces in those two countries over the coming decade, but recognizes that those countries, and the region in general, will continue to be a main focus for the U.S. military. Ongoing, albeit diminished, responsibilities in Iraq and Afghanistan, in addition to the need to respond quickly to other emerging threats in the region, result in an enduring large commitment of U.S. ground forces. In Table 4.9, 18 BCTs are indicated as the core requirement and a further 15 BCTs, included in National Command, are earmarked for CENTCOM. In like manner, there is a core requirement for 10,000 special operations forces and a further earmark of 23,000 SOF troops.

The Navy has kept one carrier strike group (CSG) in the CENTCOM area of responsibility (AOR), often surging to two. The baseline assumes a similar demand in the future. Maintaining a stable rotation

---

[4]   The approach in Figure 4.1 essentially treats costs as "fixed" and "variable," with the fixed costs corresponding to the baseline. This is a standard technique taught in business schools, although it is often desirable to go more deeply into the realm of allegedly fixed-cost expenditures to find more opportunities for tradeoffs and efficiencies—i.e., to turn fixed costs into variable costs (see the discussion in the companion monograph [Gompert, Davis, Johnson, and Long, 2008]).

Figure 4.1
Force Posture in the Analytic Baseline

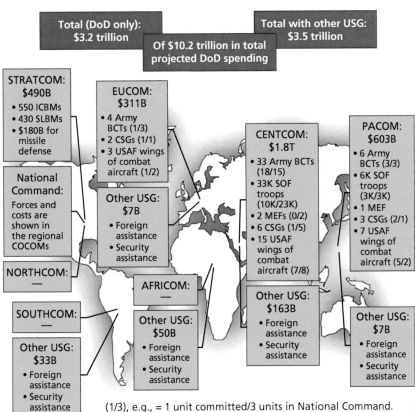

Total (DoD only):
$3.2 trillion

Total with other USG:
$3.5 trillion

Of $10.2 trillion in total
projected DoD spending

STRATCOM:
$490B
• 550 ICBMs
• 430 SLBMs
• $180B for
  missile
  defense

National
Command:
Forces and
costs are
shown in
the regional
COCOMs

NORTHCOM:
—

SOUTHCOM:
—

Other USG:
$33B
• Foreign
  assistance
• Security
  assistance

EUCOM:
$311B
• 4 Army
  BCTs (1/3)
• 2 CSGs (1/1)
• 3 USAF wings
  of combat
  aircraft (1/2)

Other USG:
$7B
• Foreign
  assistance
• Security
  assistance

AFRICOM:
—

Other USG:
$50B
• Foreign
  assistance
• Security
  assistance

CENTCOM:
$1.8T
• 33 Army BCTs
  (18/15)
• 33K SOF
  troops
  (10K/23K)
• 2 MEFs (0/2)
• 6 CSGs (1/5)
• 15 USAF
  wings of
  combat
  aircraft (7/8)

Other USG:
$163B
• Foreign
  assistance
• Security
  assistance

PACOM:
$603B
• 6 Army
  BCTs (3/3)
• 6K SOF
  troops
  (3K/3K)
• 1 MEF
• 3 CSGs (2/1)
• 7 USAF
  wings of
  combat
  aircraft (5/2)

Other USG:
$7B
• Foreign
  assistance
• Security
  assistance

(1/3), e.g., = 1 unit committed/3 units in National Command.

NOTE: We treated about two-thirds of the DoD budget as constant across strategies,
varying forces affecting only $3.2 trillion over 20 years.
RAND MG703-4.1

base to support these forces forward and maintain a surge capability
generates a total requirement of six CSGs, with one always present and
five held by National Command earmarked to COCOM contingencies but possibly available for other deployments.

The Air Force has been heavily engaged in the region since the
end of the Cold War, mounting three major combat operations there,
and the requirement to have a sizable portion of the force ready to support joint operations in the region persists. For this exercise, we have
estimated the total requirement at 13 combat wings (both fighters and

bombers), of which eight wings are with National Command and earmarked to CENTCOM.

Foreign and security assistance will continue to be important to furthering U.S. interests in the region. Current levels of U.S. aid to the region, not including most spending on Iraq and Afghanistan, are projected to continue. The 20-year total will be $163B (FY 2009$).

PACOM. The requirements for forces in the PACOM AOR remain substantial even as they are shifting. Three BCTs are still needed to maintain a rotation base for one BCT in Korea. Three other BCTs would be earmarked to PACOM in the event of the need to reinforce but are otherwise available to National Command. Islamist movements, and other sources of instability, in the Southeast Asian region generate a requirement for some 3,000 special operations forces to support operations in the region with another 3,000 available should those operations intensify.

In the event of a conflict in Korea, the bulk of ground forces would be provided by the South Koreans and the U.S. contribution would be heavily weighted toward air; five combat wings are committed to the Pacific region or along the Pacific Rim and another two are in National Command, earmarked for PACOM should combat break out. Two carrier strike groups, with another one earmarked, are available as needed to fulfill the requirement for naval strike forces. This is a lower number than was typical during the Cold War and during the 1990s. The growth in the requirement for forces of all kinds in CENTCOM has shifted the priority requirements for naval strike forces toward that area of responsibility.

Foreign aid and security assistance will continue to play a role in furthering U.S. security interests in the region. Current (modest) levels of U.S. aid to the region are projected to continue. The 20-year total will be $7B (FY 2009$).

EUCOM. The requirement for forces for EUCOM operations diminished with the dissolution of the Warsaw Pact and breakup of the Soviet Union, but the strong U.S. relationship with Western Europe and the need for continued presence throughout the Atlantic and Mediterranean will call for significant baseline forces. An Army brigade combat team is projected as a core requirement, and three further

BCTs are earmarked. Two carrier strike groups (one of which is available to National Command) and three combat air wings (two of which are available to National Command) are provided in the baseline to support NATO and respond to needs in and around the EUCOM area of responsibility.

SOUTHCOM. At present, very few forces are assigned to or primarily oriented toward SOUTHCOM. The baseline for 2008–2028 reflects this: No major active duty forces are assigned to the command, although some forces would operate intermittently in the region. Current levels of foreign aid and security assistance are projected to continue, with a 20-year total of $33B (FY 2009$).

AFRICOM. At present, very few forces are assigned to or primarily oriented toward AFRICOM, although recent years have seen consistent operations in the Trans-Sahara and the Horn of Africa (a region that will be transferred from CENTCOM's area of responsibility to AFRICOM's). The baseline for 2008–2028 assigns no major active duty forces to the command, although some forces are projected to operate intermittently in the region. Note that each alternative strategy that we examine focuses some forces on operations in that AOR. Current levels of foreign and security assistance are projected to continue, with a 20-year total of $50B (FY 2009$).

NORTHCOM. At present, very few active forces are assigned to or primarily oriented toward NORTHCOM. The baseline for 2008–2028 assigns no major active duty forces to the command, although some modest numbers of forces would engage as needed in homeland defense missions.

STRATCOM. STRATCOM manages the U.S. arsenal of strategic nuclear forces that consists of somewhat less than 6,000 accountable warheads, which will decrease over time to 3,500. These warheads are mounted on 550 ICBMs and about 430 SLBMs; some would be deployed on strategic bombers. In the baseline strategy (and, in fact, in all the illustrative strategies), this is deemed adequate to maintain global strategic deterrence against large-scale nuclear attack on the United States. The United States has also embarked on substantial efforts to develop missile defense systems. Theater missile defense and tiered national missile defense, as now being developed, aim to protect the United States and U.S. forces from an attack of modest size and by missiles of older design than those currently being developed by Russia and China. The baseline strategy accounts for this missile defense capability by projecting steady funding for the Missile Defense Agency (MDA): $9B (FY 2009$) a year for 20 years, or $180B—roughly MDA's current annual budget.

National Command. Most of the force structure is fungible. It can be put at the service of any COCOM. The baseline strategy reflects this. Although the strategy fully allocates or earmarks major active duty force structure components to the COCOMs, only a portion of those forces are considered a core command requirement and the earmarked forces are available for worldwide deployment and hence are considered part of a virtual National Command.

Even these "core" requirements could change as events dictate. The global requirement for forces will no doubt change between 2009 and 2028. The baseline strategy represents an estimate of the average requirements and available forces during that period.

Programs and Force Shifts to Support the Direct GWOT/COIN Strategy

Programs and force shifts, over and above the baseline forces, that are needed to support the Direct GWOT/COIN strategy are summarized in Table 4.10 and Figure 4.2. Only COCOMs affected by resource changes are discussed. Inherent in this strategy is the assumption that direct action by U.S. forces is necessary to combat insurgencies in those countries where they pose the greatest threat. There are a number of ways to provide the requisite increase in capabilities. The programs indicated below are judged to be able to deliver those capabilities most effectively and cost-effectively. Choosing a specific program is a critical step and demands input from both planners and programmers in the organization that uses the methodology. Further details on these programs and on their cost is available in Appendix D.

Initiatives for CENTCOM. The critical capability enhancements required for CENTCOM's contribution to the strategy are concentrated in the area of sizable, adaptable ground forces. A lesson from U.S. operations in Iraq, Afghanistan, and elsewhere is that the success of a

Table 4.10
Programs and Force Shifts for the Direct GWOT/COIN Strategy

Program/Shift	Command	Service	Cost, 2009–2028 (FY 2009 $B)
Add 6 BCTs	CENTCOM	USA	173.3
Increase USMC forces by 27,000	CENTCOM	USMC	73.4
Add 2 SOF companies	CENTCOM	Multi	0.6
Add 2 SOF companies	PACOM	Multi	0.6
Add 2 SOF companies	AFRICOM	Multi	0.6
Security assistance	CENTCOM	Other USG	36.5
Security assistance	PACOM	Other USG	6.7
Security assistance	AFRICOM	Other USG	10.0
Total (all USG)			**301.6**
Total (DoD only)			**248.4**

NOTE: Numbers may not add to totals because of rounding.

counterinsurgent strategy is strongly correlated with having adequate ground forces. This in turn implies an adequate rotation base to keep them engaged at a robust level for a long time. To this end, our illustrative program for CENTCOM adds six Army BCTs and 27,000 marines to expand the rotation base of ground forces available to DoD.[5]

Two companies of additional special operations forces are programmed to relieve the long-term strain on these forces for which there is a strong and enduring requirement in counterinsurgency operations. They would first focus on neutralizing high-value insurgent targets although, as opportunity allowed, they could work with indigenous forces to upgrade their capabilities.

---

[5] This duplicates the active duty force structure that the Bush administration intends to add to the armed forces—an increase that was *not* included in the Analytic Baseline. In January 2007, the administration proposed adding the following to the end-strength levels recommended in the 2006 *Quadrennial Defense Review Report* (QDR): 65,000 soldiers to the Army (with the principal capability increase of six Army BCTs) and 27,000 marines to the USMC.

Figure 4.2
Programs and Force Shifts for the Direct GWOT/COIN Strategy, by COCOM

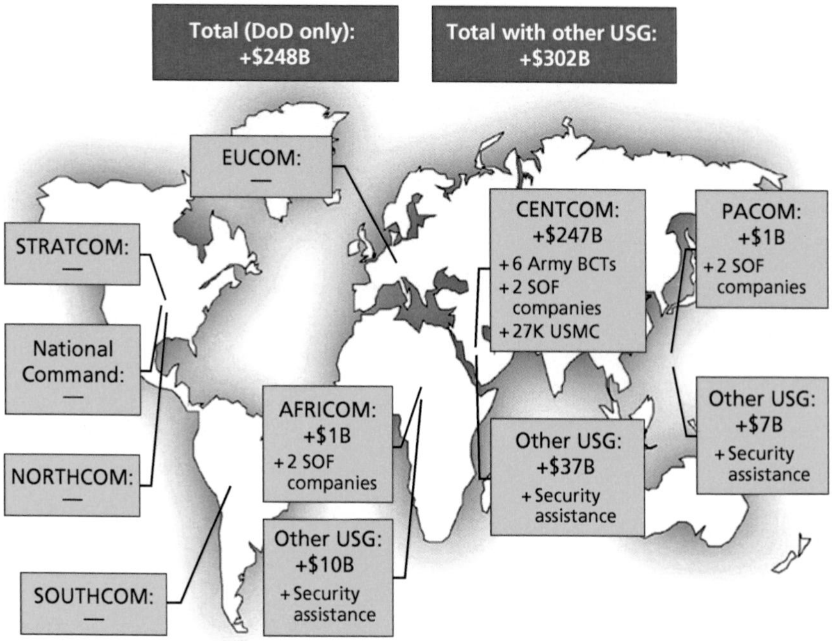

Total (DoD only):
+$248B

Total with other USG:
+$302B

EUCOM:
—

STRATCOM:
—

CENTCOM:
+$247B
+6 Army BCTs
+2 SOF companies
+27K USMC

PACOM:
+$1B
+2 SOF companies

National Command:
—

AFRICOM:
+$1B
+2 SOF companies

Other USG:
+$37B
+Security assistance

Other USG:
+$7B
+Security assistance

NORTHCOM:
—

Other USG:
+$10B
+Security assistance

SOUTHCOM:
—

NOTE: Numbers may not add to totals because of rounding.
RAND MG703-4.2

Civil capabilities are also essential to COIN operations. The State Department and U.S. Agency for International Development (USAID) would increase their presence in CENTCOM and work with U.S. military units and with partner nation civil and military authorities to improve the effectiveness of COIN efforts. Funding for this initiative is included as security assistance.

Initiatives for PACOM. Although Southeast Asia is not the primary of target of Islamist insurgency, there is a measure of Islamist insurgent activity there that demands additional measures to support this strategy.

First is strengthening special operations forces' capability in the region. Two SOF companies are added. As in the case of CENTCOM, these forces would strengthen the SOF already available to PACOM for striking high-value targets. When possible, they could complement ongoing efforts to train indigenous forces in counterinsurgency operations.

Civil capabilities are also essential to COIN operations. The State Department and USAID would increase their presence in PACOM and work with U.S. military units and with partner nation civil and military authorities to improve the effectiveness of COIN efforts. Funding for this initiative is included as security assistance.

Initiatives for EUCOM. Although EUCOM would not require additional U.S. forces under this strategy, it would be tasked to work with U.S. European allies to encourage them to develop expertise in counterinsurgency operations and to share the burden in these operations outside NATO's borders. This would include intensifying liaison and coordination with allies in ongoing COIN operations in Afghanistan and, over the longer term, in Africa.

Initiatives for AFRICOM. A key challenge for AFRICOM will be Islamist activity in North Africa, the Horn of Africa, and elsewhere. Two SOF companies would provide a capability to strike high-value targets and to cultivate a counterinsurgency capability in the command's AOR. Civil capabilities are also essential to COIN operations. The State Department and USAID would increase their presence in AFRICOM and work with U.S. military units and with partner nation civil and military authorities to improve the effectiveness of COIN efforts. Funding for this initiative is included as security assistance.

Programs and Force Shifts to Support the Build Local, Defend Global Strategy

The capabilities that need to be enhanced above those of the baseline forces and existing program to support the Build Local, Defend Global strategy are summarized in Table 4.11 and Figure 4.3. Only COCOMs affected by resource changes are discussed. There are a number of ways to provide the requisite increase in capabilities. The programs indicated below are judged to be able to deliver those capabilities effectively. Choosing specific programs is a critical step and demands input from both planners and programmers in the organization that uses the methodology.

Table 4.11
Programs and Force Shifts for the Build Local, Defend Global Strategy

Program/Shift	Command	Service	Cost, 2009–2028 (FY 2009 $B)
Cut 2 BCTs	CENTCOM	USA	−51.0
Convert CENTCOM brigade-equivalent to training and advisory units; deploy to CENTCOM and National Command	CENTCOM/CENTCOM, National Command	USA	10.5
Cut 2 BCTs	EUCOM	USA	−45.3
Move 3 BCTs from PACOM to National Command	PACOM/National Command	USA	—
Add 2 green water squadrons (GWS)	CENTCOM	USN	8.5
Add 2 GWS	PACOM	USN	6.3
Add 2 GWS	AFRICOM	Navy	6.2
Add 2 GWS	SOUTHCOM	USN	6.0
Add MALE UAV[a] squadron	CENTCOM	USAF	0.6
Add MALE UAV squadron	PACOM	USAF	0.6
Add MALE UAV squadron	CENTCOM	USN[b]	0.6
Add MALE UAV squadron	PACOM	USN	0.6
Add 2 MALE UAV squadrons	National Command	USAF	1.1
Add 2 MALE UAV squadrons	National Command	USN	1.1
Move C-17 squadron from CENTCOM to EUCOM	CENTCOM/EUCOM	USAF	—
Add long-range surveillance and strike aircraft squadron	STRATCOM	USAF	20.3
Move 6 SOF groups from CENTCOM: 4 to AFRICOM and 2 to SOUTHCOM	CENTCOM/AFRICOM, SOUTHCOM	Multi	—
Add SOF battalion	National Command	Multi	2.0
Add SOF training company	CENTCOM	Multi	0.2

Table 4.11 (continued)

Program/Shift	Command	Service	Cost, 2009–2028 (FY 2009 $B)
Add SOF training company	PACOM	Multi	0.2
Add SOF training company	AFRICOM	Multi	0.2
Add SOF training company	SOUTHCOM	Multi	0.2
Add UAV detachment (HALE)[c]	CENTCOM	USAF	1.1
Add UAV detachment (HALE)	PACOM	USAF	1.1
Add UAV detachment (HALE)	National Command	USAF	1.1
Security assistance	CENTCOM	Other USG	34.2
Security assistance	PACOM	Other USG	11.4
Security assistance	AFRICOM	Other USG	11.4
Foreign assistance	CENTCOM	Other USG	114.0
Foreign assistance	PACOM	Other USG	38.0
Foreign assistance	AFRICOM	Other USG	38.0
Total (all USG)			$219.2
Total (DoD only)			–$27.8

[a] Medium-altitude/long endurance unmanned aerial vehicle.

[b] The Navy version of the MALE UAV is imagined to be shore-based. It need not be identical to the Air Force asset, although in fact the Predator B (based on which the cost for this program was derived) has lent its basic design to a potential future Navy UAV, the Mariner. It is not a replacement for the Broad Area Maritime Surveillance Program, which is assumed to be in the baseline force.

[c] High-altitude/long endurance UAV.

Initiatives for CENTCOM. The key thrust of this strategy is to develop sizable, capable, indigenous security forces, as well as the civil apparatus—police, justice, and corrections—necessary to provide effective internal security and governance. Nowhere is the challenge more critical than in CENTCOM's AOR. If this strategy is to succeed, the large numbers of U.S. and allied forces currently in Iraq and Afghanistan must be replaced by local forces able to maintain stability and defend borders from determined troublemakers.

Figure 4.3
Programs and Force Shifts for the Build Local, Defend Global Strategy, by COCOM

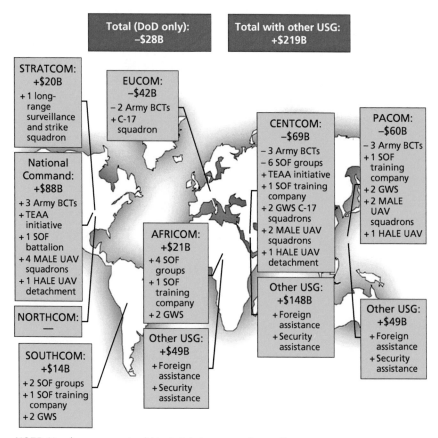

NOTE: Numbers may not add to totals because of rounding.
RAND MG703-4.3

This is a tall order. First in priority is a robust expansion of the program to train, equip, advise, and assist (TEAA) indigenous forces to prepare them to take the lead in counterinsurgency operations. An Army BCT currently in CENTCOM is re-roled to staff mobile training teams (MTTs).[6] The personnel from a BCT would provide the roughly

---

[6]   Other services could certainly make a contribution to this effort as well. For the sake of simplicity, this strategy involves only the Army.

450 new MTTs needed to cope with the continuing global threat posed by insurgency and state instability (Grissom and Ochmanek, 2008).[7] The majority of resulting units would remain oriented toward CENTCOM, but 40 percent would be placed in National Command, providing a force that can be targeted at partner nations in the greatest need. Special operations forces, for whom TEAA is already a core mission, would continue these activities, supplemented by an additional dedicated training company. As the indigenous forces take over, three brigade combat teams and six SOF group-equivalents can be expected to be freed up. The latter contingent would be divided between AFRICOM and SOUTHCOM to deal with emerging threats and enhance TEAA capabilities in those COCOMs.

Other government agencies would play a key role in this strategy; indeed, much of the additional cost comes from expenditures by the State Department, USAID, and other U.S. government agencies. Substantial funds would be devoted to building the capacity of partner governments. Under the heading of security assistance, monies would be provided to greatly enhance the ability of non-DoD agencies to dispatch civilian advisors to troubled nations. These staff, from USAID and elsewhere, would promote local capacity in areas ranging from traditional security—police, justice, and corrections—to fundamental human security—access to health care, food, and clean water. In regions in crisis, these civilians would improve the ability of U.S. agencies to work closely with the U.S. military.

An even greater amount of money would be provided as foreign assistance. The intent would be similar, but the program would have a wider mandate to promote economic development. It would be monitored by USAID officials but would not be directly administered by them in the same manner as security assistance.

The Build Local, Defend Global strategy would also include harnessing the capabilities of local forces for shallow-water naval operations (maintaining security in ports, straits, and coastal waters), as envisioned in the "thousand ship" Navy concept described by former Chief

---

[7] Grissom and Ochmanek indicate that a substantial increase in MTTs would be an important, but not sufficient, step to put the military on better footing to train foreign forces.

of Naval Operations and now Chairman of the Joint Chiefs of Staff, Admiral Mike Mullen.[8] To this end, the U.S. Navy would develop and deploy two green water squadrons. The squadrons would be organized around small ships able to operate in shallow water (see Appendix D for one possible GWS structure). The ships' scale would allow them to work in shallow, restricted waters alongside local naval forces to both complement and upgrade local capabilities through training and joint exercises.

As local forces took over the bulk of land operations, the United States would enhance its supporting role, providing ISR for those forces. The medium altitude/long endurance (MALE) squadrons available to the PACOM commander would increase by two—one operated by the Air Force and one by the Navy—to address the local forces' need for tactical-level ISR and a high altitude/long endurance (HALE) detachment operated by the Air Force would address their need for ISR on the upper end of operational-level ISR.

As the presence of U.S. ground forces deployed to the theater decreases, the requirement for strategic airlift likewise decreases. A squadron of C-17 aircraft could be freed up to meet the increase in demand for strategic airlift in EUCOM, as discussed below.

Initiatives for PACOM. For four decades, PACOM has had a strategy to provide forces to fight alongside local allied forces to underpin their defense. Up to now, the focal point for land forces has been South Korea. With its substantial growth in the past four decades, South Korea dwarfs North Korea in gross domestic product and in population, leaving it in a much improved position to defend itself against its neighbor. The United States should be able to pay increased attention to other regional threats. Other friendly forces in the region, Indonesia for example, need improved counterinsurgency capabilities. An expansion of the program is needed to help partner nations combat insurgencies. Compared with CENTCOM, the threat of insurgency within the

---

[8]   Admiral Mike Mullen described the thousand ship Navy concept in a number of speeches in 2006 (including an opinion piece in the October 29 edition of the *Honolulu Advertiser*). There would be no literal thousand ship navy but rather a network of partner nations' navies and coast guards, merchant fleets, and port operators all cooperating on common maritime security challenges.

PACOM AOR is less advanced, so a focus on preventing conditions that give rise to insurgencies and instability is warranted—additional funding would be programmed for economic development assistance and government capacity-building. This foreign assistance program would largely fall under the purview of the State Department.

Preventive measures alone are not sufficient, however. The capacity of non-DoD agencies, USAID in particular, to support ongoing COIN operations would be enhanced (this is listed as security assistance in Table 4.11). Existing forces would be complemented by an additional SOF training company, and National Command would hold a reserve of conventional MTTs ready to assist in priority areas.

These measures, along with the growing capabilities of the South Korean military, would ease the requirement that U.S. ground forces be prepared for land operations in Asia. Three BCTs could be removed from PACOM.

As local forces took over the bulk of land operations, the United States would enhance its supporting role, providing, as in the case of CENTCOM, ISR for those forces. The MALE squadrons available to the PACOM commander would increase by two—one operated by the Air Force and one by the Navy—to address local forces' need for tactical-level ISR and a HALE UAV detachment operated by the Air Force would address their need for ISR on the upper end of operational-level ISR.

Primary responsibility for shallow-water naval operations (maintaining security in ports, straits and other key SLOC points, and coastal waters) would be passed to the local forces. The U.S. Navy would develop and deploy two green water squadrons consisting primarily of ships whose scale allowed them to operate in these environments alongside local naval forces to both complement and upgrade local capabilities through training and joint exercises.

Initiatives for EUCOM. In this strategy, the United States would continue the trend of reorienting forces, especially ground forces, away from the territorial defense of Europe. The European members of NATO (and non-NATO members of the European Union) have more than enough wealth and population to defend their territory against invasion. Under this strategy, another two BCTs could be cut from the EUCOM AOR.

EUCOM would be tasked to work with U.S. European allies to encourage them to develop expertise in counterinsurgency and TEAA operations and to share the burden in these operations outside NATO's borders. This would include intensifying liaison and coordination with allies in ongoing COIN operations in Afghanistan and, over the longer term, in Africa. A C-17 squadron could be reoriented from CENTCOM to EUCOM to enable allied forces to deploy promptly to critical regions until the A400 strategic airlift aircraft enters their inventories.

Initiatives for AFRICOM. When AFRICOM comes fully on line, a key challenge will be to enable the development of capable indigenous militaries and to focus U.S. capability on striking difficult, high-value targets. Four SOF group–equivalents would be reoriented from CENTCOM to AFRICOM and a SOF training company would be added to AFRICOM. These forces would provide an ability to strike high-value targets and to cultivate a counterinsurgency capability in the command's AOR. The capacity of non-DoD agencies, USAID in particular, to support ongoing COIN operations would be enhanced. In addition, National Command would hold a reserve of conventional MTTs ready to assist in priority areas.

In an effort to prevent conditions that give rise to insurgencies and instability in the first place, additional funding would be programmed for economic development assistance and government capacity-building. This foreign assistance program would largely fall under the purview of the State Department.

Local forces would maintain the responsibility for shallow-water naval operations (defending offshore energy infrastructure and maintaining security in ports, straits, and coastal waters). The U.S. Navy would develop and deploy two green water squadrons, which would be configured around ships whose scale allowed them to operate in these environments alongside local naval forces to both complement and upgrade local capabilities through training and joint exercises.

Initiatives for SOUTHCOM. This strategy envisions that the U.S. contribution to security in the region will take the form of training local forces and providing specialized capabilities to complement local ones.

Two SOF group–equivalents would be reoriented from CENT-COM to SOUTHCOM and a SOF training company would be added to SOUTHCOM. These forces would provide an ability to strike high-value targets and to cultivate partner nations' counterinsurgency capabilities in the command's AOR.

Local forces would maintain responsibility for shallow-water naval operations (maintaining security in ports and patrolling major rivers, straits, and coastal waters). The U.S. Navy would develop and deploy two green water squadrons, which would be configured around ships whose scale allowed them to operate in these environments alongside local naval forces to both complement and upgrade local capabilities through training and joint exercises.

Initiatives for National Command. Although some U.S. forces are removed from the CENTCOM, PACOM, and EUCOM AORs, a portion of those forces are maintained by the centrally managed National Command.

Implementation of this strategy reduces the requirement for eight forward-deployed BCTs. Of these, three BCTs are shifted to National Command to hedge against misjudgment in one of the theaters and the need to refocus ground forces on a requirement that could emerge from an AOR (most likely CENTCOM's). A further BCT, as mentioned above, would be re-roled to staff military training and advisory units. The preponderance of these units would be deployed to CENTCOM, but 40 percent would stay in National Command, where they would be available to bolster whichever COCOM's TEAA needs seemed most pressing.

In addition, one SOF battalion is held by National Command to strike in a regional command if needed. Four squadrons of MALE UAVs and a detachment of HALE UAVs provide surge capability to both monitor and strike emerging threats.

Initiatives for STRATCOM. With the United States "offshore," supporting allied and partner forces, a squadron of long-range surveillance and strike aircraft is added to STRATCOM to provide the ability to strike anywhere on the globe promptly and with precision.

Programs and Force Shifts to Support the Respond to Rising China Strategy

The capabilities that need to be enhanced above those of the baseline forces and existing program to support the Respond to Rising China strategy are summarized in Table 4.12 and Figure 4.4. Only COCOMs affected by resource changes are discussed. There are a number of ways to provide the requisite increase in capabilities. The programs indicated below are judged to be able to deliver those capabilities most effectively, and cost-effectively. Choosing specific programs is a critical step and demands input from both planners and programmers in the organization that uses the methodology.

Initiatives for CENTCOM. The focus of this strategy shifts from the CENTCOM region and toward a substantial scaling back of U.S. participation on land. Three BCTs are removed from CENTCOM orientation. That said, there is still concern about the threat posed by terrorist havens. Substantial forces remain in the baseline to provide for CENTCOM requirements. In addition, security and foreign assistance to the area is increased. This aid effort shares a goal with that of the Build Local, Defend Global strategy—reduce the burden on the U.S. military by building local capacity—but it is conducted at a much reduced scale.

The United States will also help the countries in the region ensure the security of their ports, of the energy infrastructure in coastal regions and offshore, and of the SLOC choke points. To this end, a green water squadron of U.S. ships appropriate to operations in these restricted waters will be developed for use in the CENTCOM AOR.

Initiatives for PACOM. It is important to note that the strategic thrust of the Respond to Rising China strategy, as described above, is to cultivate a constructive relationship with China. This approach does not have direct resource implications, but it informs initiatives in PACOM. Although significant military forces are added to U.S. regional baseline capabilities, the aim is to hedge against, not provoke, an arms race and maintain an advantage in military power.

The U.S. Navy in particular receives substantial new assets. Twenty-four capital ships will be added to the Pacific Fleet, giving a

Table 4.12
Programs and Force Shifts for the Respond to Rising China Strategy

Program/Shift	Command	Service	Cost, 2009–2028 (FY 2009 $B)
Cut 3 BCTs	CENTCOM	USA	–75
Move 3 BCTs from EUCOM to National Command	EUCOM/National Command	USA	—
Add 2 GWSs	PACOM	USN	8.5
Add GWS	CENTCOM	USN	3.3
Add GWS	AFRICOM	USN	3.1
Add GWS	SOUTHCOM	USN	3.0
Add 2 GWSs	National Command	USN	5.7
Add 4 SSGNs (Submersible, Ship, Guided, Nuclear); a nuclear-powered cruise-missile submarine	PACOM	USN	4.2
Add 4 CG(X)s (a future cruiser)	STRATCOM	USN	17.5
Move 12 DDG-1000/CG(X)s from EUCOM to PACOM	EUCOM/PACOM	USN	—
Add 12 DDG-1000/CG(X)s	PACOM	USN	49.3
Add medium-range bomber wing	PACOM	USAF	61.1
Add long-range surveillance and strike squadron	STRATCOM	USAF	20.3
Add long-range conventional missiles	STRATCOM	USN	1.0
Add HALE UAV squadron	STRATCOM	USAF	5.1
Enhance national missile defense	STRATCOM	USAF	81.8
Add HALE UAV detachment	PACOM	USAF	1.0
Add HALE UAV detachment	STRATCOM	USAF	1.0
Security assistance	CENTCOM	Other USG	19.0
Security assistance	PACOM	Other USG	2.9
Security assistance	AFRICOM	Other USG	2.9
Foreign assistance	CENTCOM	Other USG	20.9
Foreign assistance	PACOM	Other USG	7.9
Foreign assistance	AFRICOM	Other USG	9.5
Total (all USG)			$253.7
Total (DoD only)			$191.0

Figure 4.4
Programs and Force Shifts for the Respond to Rising China Strategy, by COCOM

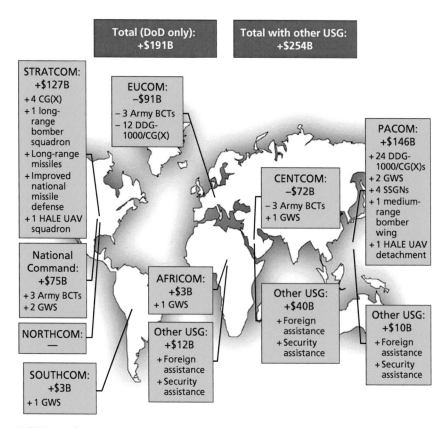

NOTE: Numbers may not add to totals because of rounding.
RAND MG703-4.4

decisive regional blue-water advantage and providing a robust capacity to operate in any threat environment.

The Navy would also develop and deploy two green water squadrons. These would be configured around ships small enough to operate in these environments alongside local naval forces to both complement and upgrade local capabilities through training and joint exercises. These units would provide a basis for naval cooperation with China as well, perhaps in securing SLOCs. In the event of hostilities, however, they would also enhance U.S. littoral warfare capability in such key capacities as antisubmarine and antimine warfare.

In addition, four SSBNs converted to SSGN conventional cruise missile carriers and SOF platforms will be reoriented to PACOM to maintain a threat to targets in and around China while preserving a high degree of stealth and survivability.

A detachment of HALE unmanned aerial vehicles is brought into the force to enhance U.S. ability to develop operational-level ISR in and around China.

To complement the short-range tactical fighters already in the baseline, a wing of medium-range bombers will be added to PACOM. These aircraft will provide enhanced capability to threaten targets along the Chinese littoral from outside the range of most of China's ballistic missiles.

Security and foreign assistance round out the capability enhancements applied to PACOM. Although modest relative to the sums in the Build Local, Defend Global strategy, new resources carefully targeted at improving the capacity of regional governments and at promoting economic development will improve U.S. standing in the region.

Initiatives for EUCOM. The center of gravity of this strategy shifts solidly to the Pacific Rim. The result is that 12 capital ships are transferred from the Atlantic Fleet to the Pacific Fleet. Moreover, three BCTs are shifted from EUCOM to National Command for allocation as needed to unforeseen requirements. Although no military capabilities are assigned for this purpose, the United States should engage with European allies to encourage a broader European role in CENTCOM, AFRICOM, and elsewhere while the United States focuses on China.

Initiatives for AFRICOM. DoD would look to AFRICOM to limit Chinese influence in Africa. A key element would be to deepen military ties. The U.S. Navy would develop and deploy a green water squadron to work with African countries on defending coastal waters, to include offshore and nearshore energy infrastructure. The squadron would be configured around ships small enough to operate in these environments alongside local naval forces. It would train and exercise with a local country's navy and then move on to another country. A cycle of approximately six weeks per country would allow the U.S. Navy to maintain contact with eight or nine countries per year.

To help match increased Chinese influence in Africa, the United States would also increase development assistance to worthy nations. This program would largely fall under the purview of the State Department.

Initiatives for SOUTHCOM. The DoD would also look to SOUTHCOM to limit Chinese influence. A key element would be working with South American countries to defend their coastal waters and execute riverine operations. The U.S. Navy would develop and deploy a green water squadron. It would be configured around ships small enough to operate in these environments alongside local naval forces. It could train and exercise with a local country's navy and then move on to another country.

Initiatives for National Command. Three of the six BCTs removed from CENTCOM and EUCOM are maintained in the force structure and put into the same pool for allocation to whatever COCOM has an unanticipated requirement.

In addition, two more green water squadrons are developed and allocated to National Command so that operations in any COCOM can be readily expanded as needed.

Initiatives for STRATCOM. China is expanding the range and accuracy of its conventional strike systems. It is also upgrading and extending the range of its ISR capacity. This strategy envisions that STRATCOM will assume responsibility for longer-range aircraft that can strike targets on the littoral of China promptly from a long distance away. There are two STRATCOM strike programs: a squadron of long-range surveillance and strike aircraft[9] and long-range conventional ballistic missiles. An additional squadron of the Air Force's planned long-range surveillance and strike aircraft provides increased numbers of stealthy aircraft that could fly from CONUS and penetrate a robust air defense network. Long-range conventional ballistic missiles (whether

---

[9]   In considering future programs, the Air Force frequently considers modernizing the existing bomber fleet and refers to the option alluded to here as a "future long-range bomber." We prefer the term "long-range surveillance and strike aircraft," which explicitly names what should be two core capabilities for a stealthy future platform.

developed for submarines or land-based silos) provide a prompt strike option when time or distance precludes the use of aircraft.

With the risk of China expanding its ICBM capability to include, over time, mobile land-based and SSBN-based systems, investment in ballistic-missile defense is expanded. Four new CG(X)s are programmed to provide enhanced theater missile defense. A boost-phase interceptor system will also be developed to cope with higher-velocity, solid rocket fueled systems. Although this system will not be impenetrable in the long run, it provides a measure of protection against emerging Chinese survivable systems in the short run and the foundation for more robust protection in the future. It is not intended to replace the overwhelming advantage the United States has in numbers of survivable, accurate reentry vehicles.

A squadron of HALE unmanned aerial vehicles is brought into the force to enhance the U.S. ability to develop operational-level ISR in and around China and in other areas of interest.

Finally, China has demonstrated its ability (and perhaps signaled its willingness) to shoot down low earth orbit satellites. This is precisely the type of satellite on which the United States depends for theaterwide ISR. Communications and global positioning system (GPS) satellites could also be at risk. Therefore, as a backup capability, i.e., to plug a gap in coverage rapidly should the need arise, a program that procures additional HALE UAVs is introduced. These would be held back in time of crisis specifically to ameliorate any loss of satellite ISR coverage; they could also serve in a limited capacity as theater communications relays.

Alternative Expressions of Costs

Principles

Basic methods for estimating defense costs were developed decades ago and described in classic books on defense economics (Hitch and McKean, 1965) and more specifically on costs (Fisher, 1971). Since practice often falls short of the standards proposed in those earlier years, it is necessary from time to time to rediscover the principles and enforce their application.[10] This is necessary because it is difficult to do costing work well. It is often not in the interest of those advocating programs to reveal the full anticipated costs, and Congress focuses largely on the current budget. In addition, some considerations are new.

The economic principles that we suggest in our methodology, although intended to be relatively simple, are those of the following list. The first three are familiar, even if not applied consistently. The fourth is known to some and widely ignored. The remainder are more unusual.[11]

1. *Constant dollars.* Using inflation-corrected figures costs.
2. *Life-cycle costs.* Estimating complete life-cycle system costs, particularly the costs of R&D, acquisition (including fielding the forces acquired), and subsequent O&S.
3. *Time streams.* Paying attention to the time stream of expenditures because of the need to stabilize the DoD budget rather than imagining that Congress will nicely tolerate large year-by-year fluctuations as big programs come in or conclude.
4. *Recognizing the value of money.* Characterizing the future economic consequences of the principles outlined above by reporting the net present value (NPV) of future obligations.[12]
5. *Including deferred expenses.* Reflecting through an "effective burden factor" the eventual DoD-related costs that will be borne by other agencies for retirement and retirement health plans, for example.[13] This is sometimes referred to as accrual accounting.

---

[10] Many defense analysts look to the Congressional Budget Office (CBO) for consistently good reporting practices and data on defense programs.

[11] Even this list is incomplete. For example, it is sometimes important to keep track of capital costs, labor costs, and depreciated costs because they may be subject to special constraints or may represent special concerns needing attention.

[12] Throughout this monograph, we refer to net present value, but we are referring to costs, which are sometimes expressed as negative NPVs.

[13] The burden may be seen as a virtual tax on DoD expenditures, one generating revenues for a "trust fund" to be used when the obligations come due. No such trust fund exists and

6. *Reporting all costs, i.e., all government costs, to the taxpayer.* Reporting total U.S. government costs where relevant, not just those costs borne by DoD itself (e.g., include anticipated costs for the State Department).

7. *Reporting by both service and COCOM.* Developing special reports suitable for different audiences, with these to include—consistent with the spirit of this monograph's emphasis on the COCOMs as operating units—breakdowns by COCOM as well as by service, even though the actual budgets belong to the services.

8. *Extraordinary costs.* Allowing for the extraordinary (noncore) costs associated with actual wars, prolonged crises, or prolonged activities such as counterinsurgency. These are particularly important and are taken up below.

Total Costs of Strategy

Let us now illustrate the principles with aggregate-level calculations that spare the reader from many details but allow us to extract insights. What follows is an unfolding story, starting with relatively mundane information and then adding both sophistication and important components of cost.

DoD Costs. Figure 4.5 shows constant-dollar expenditures over time for the alternative strategies. Figure 4.6 shows cumulative constant-dollar expenditures over time and is easier to follow. Both figures use data based on life-cycle costs.

Looking at these figures, we see that the Analytic Baseline strategy, by definition, costs nothing extra. The Direct GWOT/COIN strategy costs about $12B more a year, primarily because of the additional ground forces that are added to the structure permanently. The annual cost of the Respond to Rising China strategy rises sharply to account for initial capital investments but ultimately costs almost $40B less over 20 years than the Direct GWOT/COIN strategy. The Build

---

the government merely pays the bills when the time comes, but the artifice is sound economics because it clarifies the implications of future obligations.

Figure 4.5
DoD Expenditures over Time, by Strategy (Relative to the Baseline)

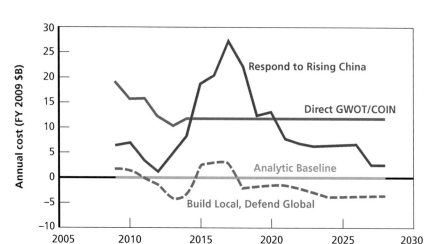

NOTE: The costs of extraordinary (noncore) operations, such as for war, are not included.

RAND MG703-4.5

Local, Defend Global strategy costs DoD much less than the other alternatives from the outset.

Net Present Value. Economists prefer NPV calculations because money that need not be spent now does not need to be borrowed or extracted from the economy by taxation. Figure 4.7 shows how the cumulative obligations in NPV terms build as we look further into the future in accounting for future obligations. Following the procedure mandated for agencies by the Office of Management and Budget (OMB), the figure assumes a real discount rate of 3 percent (about 5 percent before inflation) (Office of Management and Budget, 2006). The primary point of Figure 4.7 is to show that, for a low discount rate such as 3 percent, the NPV of future obligations can continue to grow for a long time, although the curves flatten out eventually. This is particularly true for the Direct GWOT/COIN strategy because the additional force structure is assumed permanent.

NPV calculations are sometimes done with a shorter horizon, such as 20 years, and are sometimes based on different discount rates. Over

Figure 4.6
Cumulative DoD Expenditures over Time, by Strategy

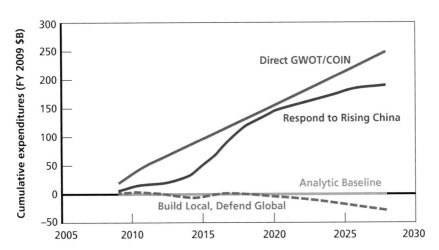

NOTE: The costs of extraordinary (noncore) operations, such as for war, are not included.

RAND MG703-4.6

the decades, OMB's guidance on what the real discount rate "should" be has varied by more than a factor of two; the current guidance (3 percent) is low and OMB's more general guidance is that calculations should be done for real discount rates of both 3 percent and 7 percent—the latter being an estimate of the long-term pretax gain from private investment in the U.S. economy (Office of Management and Budget, 2003). Economists disagree on the discount rate that should be used by government agencies and recognize that the appropriate rate depends on the agency. Market-oriented economists tend to prefer higher discount rates, arguing that when the government taxes or borrows, it drains money from the economy and can lower growth. Other economists favor lower discount rates because, especially in the present era, there is great concern about passing a heavy burden of debt on to future generations. Someone primarily concerned about trimming the near-term budget tends to favor high discount rates because doing so creates a strong argument for deferring expensive modernization (expenditures later are less painful than expenditures now). Those

Figure 4.7
Net Present Value of DoD Obligations as a Function of Horizon

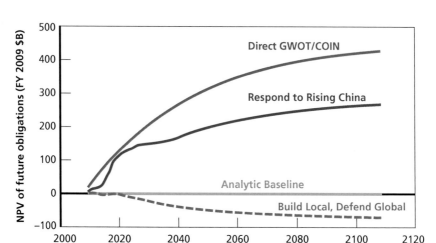

NOTES: NPV assumes a discount rate of 3 percent. The incremental costs of extraordinary (noncore) operations, such as for war, are not included.
RAND MG703-4.7

eager for immediate change, who may also see great but unprovable economic benefit in modernization, will therefore tend to favor lower discount rates.[14]

As we show below, the discount-rate assumption has a large effect on the relative net present value costs of the strategies.

Table 4.13 summarizes cost comparisons for DoD expenditures of the sort that are usually considered in defense planning. We include the widely used FYDP and 20-year costs, but the net present value of future obligations is a better measure.

DoD-Related Deferred Expenses. One concern of economists for many years was that the true cost of defense was being underestimated because of various deferred expenses that would eventually be paid by other agencies for pensions, health care, and other matters. At one time, this was a large shortcoming of the accounting system.

---

[14] An interesting discussion of the different philosophical arguments, along with citations to the original literature, can be found in a joint effort of the American Enterprise Institute and Brookings Institution (Sunstein and Rowell, 2005).

Table 4.13
Economic Comparisons of Strategies (DoD Costs in FY 2009 $B)

			NPV Horizon	
	FYDP	20-Year	20-Year	Forever
Total DoD projected spending	3,050	10,167	7,791	16,546
Strategy				
Analytical Baseline	0	0	0	0
Direct GWOT/COIN	84	248	194	425
Build Local, Defend Global	–6	–28	–18	–69
Respond to Rising China	32	191	148	267

NOTES: The costs of extraordinary (noncore) operations, such as for war, are not included. NPV calculations assume a real discount rate of 3 percent and consider obligations over both a 20-year horizon and into the indefinite future.

Most of these problems have been eliminated as the result of changes made in the 1980s and even quite recently. There continue to be some unaccounted-for obligations, but they are relatively small as a fraction of the budget and are not obviously very different across the strategies that we consider, although the manpower intensity of the Direct GWOT/COIN strategy might set it apart from the others. Such costs will not be discussed further here because they are not essential to any discussion of alternative national strategies.[15] Not surprisingly, they are largely ignored by decisionmakers except for those charged with related responsibilities.

Governmentwide Costs. One of the biggest corrections to normal DoD costing of national strategies should be to include the anticipated expense to the State Department and other agencies where success of a given strategy depends on those agencies' effective contributions. This is strongly the case for the Build Local, Defend Global strategy, as Figure 4.8 illustrates. Governmentwide costing adds about $13B a year to the cost of the strategy, resulting in a roughly $240B difference in the 20-year cost of the strategy, compared with DoD-only costs.

---

[15] CBO publications discuss such matters (Congressional Budget Office, 2007b). See also a paper from the Harvard Law School (Kohyama and Quick, 2006).

Figure 4.8
DoD Versus Governmentwide Costs

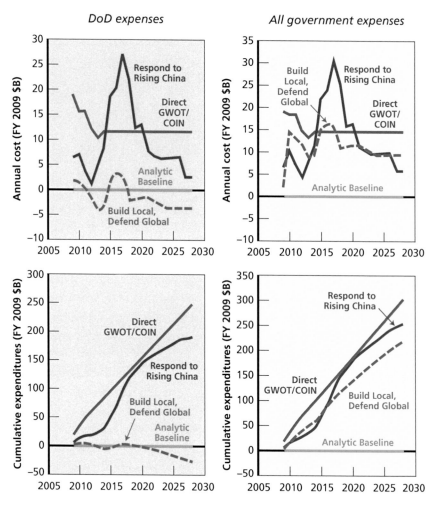

NOTE: The costs of extraordinary (noncore) operations, such as for war, are
not included.

RAND MG703-4.8

   To DoD readers at this point, the obvious question may be whether
the Build Local, Defend Global strategy would mean that DoD would
"send money to the State Department" or to other agencies. On a relative basis, the answer is yes. However, it is a matter of where additions

to the baseline budget go, not a matter of trading baseline structure for a program of foreign assistance. The bigger concerns should be the merits of the various strategies and the feasibility of greatly increasing the State Department's foreign assistance budget and its infrastructure for using such funds well. The strategy depends on such actions taking place.

Sensitivities to Discount Rate and the Convention for Calculating NPV

We have mentioned the superiority, from the viewpoint of economic theory, of using NPV calculations. As noted above, however, there are different ways to make such calculations. Figure 4.9 shows the consequence of using 7 percent rather than 3 percent as a discount rate and of accounting for expenditures into the indefinite future rather than over the more limited horizon of 20 years as is sometimes done. We see that the costs of the strategies change greatly with the longer horizon and significantly with the discount rate as well. Still, in all cases, the

Figure 4.9
Cost Comparisons in Net Present Value Terms: All USG Expenses

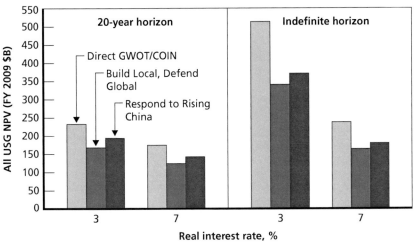

Direct GWOT/COIN strategy, which adds the greatest amount of new force structure, is substantially more expensive than the others.

The Special Costs of War or Other Intensive Operations

As noted in Chapter Three, which laid out the anticipated flow of analysis, it is traditional for U.S. defense planning to focus on "core" defense costs, which do not include the extraordinary costs of intensive military operations in war or something like the continued counterinsurgency campaign that is now ongoing in Iraq and Afghanistan. To ignore them as merely speculative would be bizarre in the current era, however. Further, the strategies we are comparing differ substantially in the extent to which such extraordinary costs are to be expected. It is part of the character and premise of the Direct GWOT/COIN strategy to pursue a highly proactive intervention-intensive approach in the Middle East. To be sure, good fortune or early successes might make such efforts unnecessary in the long term. So also, it might be that *failure* to mount such efforts in the other strategies would lead to unavoidable and expensive conflicts. However, although any of the strategies might fail and lead to high-expense conflicts or adaptations, only the Direct GWOT/COIN strategy seems almost to imply a lengthy period of noncore costs for operations. It would therefore be disingenuous not to acknowledge this. Table 4.14 does this by adding to the costs of the Direct GWOT/COIN strategy a rough estimate of the noncore costs. They are merely illustrative and highly uncertain, but they make the point that—if included—they dominate the comparison.

For illustration we used an average annual cost of $85B, which could be rationalized in various ways.[16] We assumed success in 20 years, after which no further special operations peculiar to the GWOT/

---

[16] The operating costs of operations in Iraq and Afghanistan are projected as roughly $2,000B, after some exclusions. Assuming something similar every 20 years implies about $85B a year. The estimate's rough magnitude has an objective basis (Orszag, 2007). Moreover, such calculations are relatively straightforward given determination to do them. Some were made before the war, generating results that were relatively prescient (Nordhaus, 2002; Congressional Budget Office, 2002). This said, some economists argued that the cost of war would be less than the cost of containment (Davis, Murphy, and Topel, 2003); they still believed that as of 2006 (Davis, Murphy, and Topel, 2006).

Table 4.14
USG Cost Comparisons with Predictable Special Costs Included ($ Billions)

Strategy	FYDP Costs			% of 20-Year Cost of Operation Outside Core Budget			NPV Costs		
	Excluded	Included		Excluded	Included		Excluded	Included	
		50	100		50	100		50	100
Analytic Baseline	0	0	0	0	0	0	0	0	0
Direct GWOT/ COIN	98	360	608	302	1,172	2,002	513	1,180	1,816
Build Local, Defend Global	59			219			342		
Respond to Rising China	48			254			371		

NOTE: NPV calculations assume a 3 percent real discount rate and an indefinite horizon.

COIN strategy would be needed. Table 4.14 includes columns based on that estimate and one half as large (100 percent and 50 percent, respectively). The cost associated with noncore operations dominates the calculations. We show the additional costs only for the Direct GWOT/COIN strategy because it is unique in virtually implying the need for significant, continued intervention. *All* of the strategies would likely incur some costs associated with conflict, and any of the strategies could fail, resulting in much greater-than-predicted costs to the United States. Those supporting the Direct GWOT/COIN strategy might argue that failure to adopt that strategy would lead to disaster, including an even larger conflict. What happens if strategy fails is the subject of the next section.

Costs When Strategies Fail

As discussed in Chapter Three, strategic analysis should consider how a candidate strategy would or would not lay the basis for major adaptations in the future if, in effect, the strategy should fail. Further, well-formed strategies should also include operational and strategic hedges with the potential need for adaptations in mind. It follows that the potential cost of adaptations might be estimated for use in comparing strategies.

We have not estimated such costs with any care in this monograph, for lack of time, but we offer some speculations. In the Build Local, Defend Global strategy, the intent is to work closely with traditional allies and partners (NATO, Australia, Japan, and others), while at the same time working diligently with partners in the Muslim world and elsewhere to develop capable local security forces. Those local forces would be on the front line of meeting security challenges. However, the United States might well have to intervene if the local forces were overwhelmed (as they might be if the local government proved excessively corrupt, inept, or unpopular). What might intervention entail economically?

Where feasible, the United States would seek to follow the Desert Storm script—lining up allies, having them contribute forces or cash, and sharing the cost of U.S. actions taken for the benefit of all. In such a case, costs might be quite low. Because of foreign contributions, DoD may even have "made a profit" on the first Iraq war, although full national costs were not fully covered (Johnson, 1991). That, of course, would be a favorable case. In a less-favorable case, intervention might be more unilateral and without sizable contributions from either allies or oil revenues. Costs might then be comparable to those of the Afghan operation. Were we to average the two possibilities as equally plausible, the cost of adaptation might be estimated to be on the order of $500B over 20 years. Another approach would be to assume that, if the strategy failed, a shift to the Direct GWOT/COIN strategy would be called for, in which case costs would be as high as shown in Table 4.14 and Figure 4.10.

If the Respond to Rising China strategy failed, there might be at least two causes. First, its minimalist approach to the Middle East might prove to have been unwise. Although the strategy hedges by retaining enough force structure and forward presence to protect critical U.S. interests such as the continued flow of oil, it is possible that Islamist fervor would lead to revolutions and chaos in the region, requiring

Figure 4.10
Implications of Including Costs of Extraordinary Operations

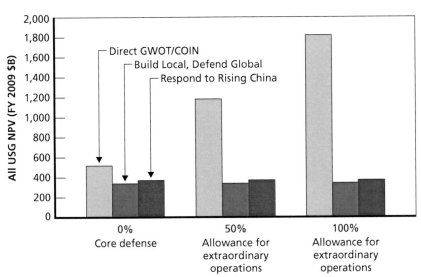

RAND *MG703-4.10*

heavy U.S. intervention. The costs of that could be as high or higher than those that the United States has been recently experiencing (perhaps $1,000B–$2,000B over 20 years). Another way that the strategy could "fail" would be if China chose to intensify the competition and seek outright military superiority in East Asia. Such a situation would be potentially open-ended, so we have not bothered to estimate its costs. On a more limited basis, one can imagine a conflict over Taiwan involving a short but intense conflict with limited supplemental operating costs but a substantial bill for replacing and modernizing air and naval forces after wartime losses or demonstration of obsoleteness. Costs could easily be on the order of $100B or even much more.

In summary, it is possible to estimate the potential cost of major strategic adaptations, but doing so is speculative. Its value is primarily in highlighting the issues and buttressing appreciation for the need to craft well-hedged strategies that would leave no vacuums and permit adaptations if they proved necessary.

Other Cost Breakdowns, by Service, COCOM, and Combinations

In this chapter, we have highlighted a number of ways to characterize the costs of strategy. Some of those are of interest when taking a careful look at the national economics involved but are less of interest when DoD offices are preparing budgets or reporting those budgets to the military departments and Congress. This is particularly so because Congress provides funding through the authorization accounts of the military services and defense agencies. Appendix C describes the simple tool and database that we used to generate different kinds of cost reports, such as reports showing the consequences to each service and COCOM of each strategy over a particular period of years and with a particular kind of costing (e.g., constant dollars or net present value).

Noneconomic Costs

We have not attempted in this monograph to make separate comparisons of the noneconomic costs of the several strategies. The comparisons are well worth making, however, because the strategies are quite different in character. As discussed above, some of these costs are misnamed, with the term "constraints" perhaps being more apt. The Direct GWOT/COIN strategy would emphasize larger ground forces and manpower-intensive operations—with the potential for continued stress of the services' ability to recruit, retain, and maintain quality. The Build Local, Defend Global strategy assumes a massive U.S. effort to increase foreign assistance relevant to counterinsurgency, something that would require creating or recreating capabilities that are today in extremely short supply. It is unclear whether political or organizational constraints could be overcome with effective results. The Respond to Rising China strategy, which requires substantial capital investments, would face constraints in the defense industrial base and service procurement budgets.

CHAPTER FIVE

Integrated Portfolio Analysis of Illustrative Strategies

Basic Concepts

In this chapter, we use a portfolio-management approach to characterize strategies' expected consequences, risks, and costs (a fuller treatment would also include upside potential). Further, we use a simplified version of exploratory analysis to understand how the answers change with differences in assumptions or judgments. Doing so is essential because uncertainties are large in many cases, as are differences in perceptions and judgments.

We begin by relating portfolio management to the larger theme of capabilities-based planning.

Capabilities-Based Planning and Portfolio Management

Capabilities-based planning (CBP) is now a cornerstone of DoD thinking.[1] A definition that suits the context of this monograph is

> Capabilities-based planning is planning, under uncertainty, to provide capabilities suitable for a wide range of modern-

---

[1] CBP was introduced to DoD in 2001 and reinforced by implementation activities (Rumsfeld, 2001; Joint Defense Capabilities Study Team, 2004; Rumsfeld, 2006).

day challenges and circumstances while working within an economic framework that necessitates choice.[2]

This definition reminds us that the Department of Defense is responsible for pursuing multiple objectives simultaneously—across the globe and in a vast range of circumstances, some of them unforeseeable. DoD must do so even with resources that are distinctly limited, although huge by the standards of other countries. CBP is not a blank check for addressing shortfalls but a disciplined approach to making choices under uncertainty while working within (and informing decisions about) budgets. Although it is frequently mischaracterized on this score, CBP includes threat-based analysis but avoids obsession with "point scenarios" that obfuscate uncertainty.

Because of this need to pursue multiple goals and operational objectives, in multiple places with multiple challenges, and in multiple potential circumstances, and to do so for the near, mid, and long terms, it is useful to think in terms of portfolio management. Rather as personal investors have a portfolio of stocks, bonds, real estate, and other instruments to satisfy different objectives and needs (e.g., long-term capital gain, current income, and risk mitigation), so also DoD invests in a mix of instruments for its multiple objectives. And, pursuing the same analogy, DoD must routinely rebalance its portfolio: increasing its emphasis on some activities and instruments while decreasing the emphasis on others and disinvesting in still others. In doing so, it must routinely make choices, including about how to manage risks. This is by no means the creature of a particular administration. The basic concerns have been on the minds of defense secretaries since at least the early 1960s.[3]

---

[2]   This formulation and its underlying theory (Davis, 2002) have been used in numerous contexts (National Research Council, 2005; Technical Cooperation Program, 2004; Fitzsimmons, 2007; Chairman of the Joint Chiefs of Staff, 2007). Ironically, CBP can be regarded as a mere expression of common sense (Fitzsimmons, 2007).

[3]   The continuity and evolution of such concerns are described elsewhere (Davis, 1994c; Chu and Berstein, 2003; Johnson, Libicki, and Treverton, 2003). The classic book on systems analysis (Enthoven and Smith, 1971) was reissued in 2005 with new introductory material by defense undersecretaries Kenneth Krieg and David Chu, noting this continuity across

Portfolio-management methods are remarkably general, even with defense applications specifically. They can be applied at different organizational levels and to different classes of problem.[4]

Methods for Portfolio Analysis

The portfolio methods that we recommend for defense department use are rather different from those used by Wall Street in developing investment portfolios.[5] Financial investors can draw on decades of empirical information about fluctuations in business cycles and stock market values and on similarly rich information about past engineering developments, among others. Much can be done to model risk quantitatively, to assess the relative risks of alternative portfolios, and even to assess the value of investing in high-risk, high-payoff activities.[6] DoD, in contrast, cannot typically balance failures in one region of the world by making special gains in another. Nor can it measure degrees of risk with any precision. Instead, it uses methods such as testing force structure and posture against defense-planning scenarios and historical information on the frequency of past types of crises and conflicts,[7] or using balance assessments and judgment informed by COCOMs, among others.

The methods that we have found especially useful in portfolio management–style thinking in strategic planning for defense involve

---

periods and administrations. Risk-management issues have been explicitly emphasized in recent years (Rumsfeld, 2001).

[4] Published studies illustrate this diversity for weapon-systems acquisition (Davis, Shaver, and Beck, 2008) and strategic planning (Davis, Gompert, and Kugler, 1996).

[5] Portfolio theory in finance and business is described in a classic (Markowitz, 1952) and numerous more recent sources (Elton et al., 2006; Hagstrom, 1999; Swisher and Kasten, 2005).

[6] One author of this monograph (Davis) benefited from conversations with Scott Matthews on Boeing's use of "real-options" theory as extended in ways that have made it more practical (Datar and Matthews, 2004). There are remarkable relationships to defense planning, despite differences.

[7] Such methods are used routinely in DoD analysis by the Joint Staff's J-8 and OSD's PA&E. They are part of DoD's Analytic Agenda. The earliest version was probably due to work championed by MG Mark Hamilton in J-8 in the mid-1990s.

(1) scorecards, (2) use of a related software instrument, RAND's portfolio analysis tool (PAT), and (3) exploratory analysis using PAT.

Scorecard Methods. Scorecard methods for comparing alternatives are by now familiar, whether in everyday consumer purchasing or in Pentagon briefings. A cognitively effective format is the familiar colored "stoplight chart" with options in rows and measures of how well the options are expected to perform (their expected "goodness"), as illustrated schematically in Figure 5.1 with options A, B, and C, assessed by measures M1, M2, and M3. The usual convention is that red is bad, green is good, and yellow is mediocre or marginal. Orange and light green can be used for in-between evaluations. For the convenience of those reading a gray-scale hardcopy, colors are indicated by letters R, O, Y, LG, and G.[8]

Such scoreboards allow decisionmakers to see simultaneously how options fare in each category of goodness that they care about. In the context of this monograph, the categories of goodness might relate to projected health of the relevant "operating units" (COCOMs) in the

Figure 5.1
Schematic Top-Level Scorecard

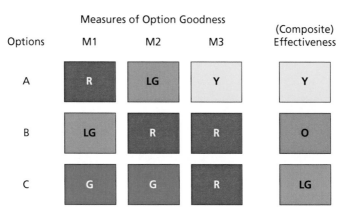

RAND *MG703-5.1*

---

[8]   Scorecard methods in policy analysis were developed 30 or more years ago (Goeller et al., 1977). The business literature now includes simplified versions (Kaplan and Norton, 1996).

near, mid, and long terms. Health could be assessed in terms of capability for test-case challenges and other operating responsibilities.

The usual scorecards are notoriously unsatisfactory because the decisionmaker has little basis for believing the colors. It is important to be able to "zoom" or "drill down" to understand the basis of top-level judgments. Senior decisionmakers have very limited time available for reviews, but they should be able—even if only for spot-checking— to do such drill-downs. If they demand that studies to support them are structured to permit this, the resulting analysis will come to be more systematic and rigorous. Reviews can then be more efficient and effective.[9]

The Portfolio Analysis Tool. RAND's Portfolio Analysis Tool is very useful for structuring analysis in the first place, for analyzing the consequences of many changes of assumption, for communication, and for summarizing results in a scorecard format as well as various charts (see also Appendix B).

One function of PAT is to generate straightforward multicriteria comparisons of effectiveness and cost. It can also generate "cost-effectiveness" comparisons using single-number measures of both cost and effectiveness, although that is fraught with danger for reasons indicated below and should be done only in the context of exploratory analysis.

Exploratory Analysis. Exploratory analysis examines outcomes for a wide range of assumptions, varying those assumptions *simultaneously*, rather than one at a time, while holding other things constant. By doing so over the full range of plausible values for the key assumptions, analysis identifies what combinations of assumption lead to good, bad, or marginal results. This is very different from working all the details of a single test scenario with a single set of assumptions and then making a few excursions. The results may vary a good deal but, in favorable cases, some robust conclusions emerge. Exploratory analysis in the con-

---

[9] Such considerations motivated a 2005 request by the Under Secretary of Defense for Acquisition, Technology, and Logistics that RAND develop a generic version of its portfolio analysis methods (Davis, Shaver, and Beck, 2008), which we have adapted for this study.

text of portfolio analysis using PAT has a number of new and unusual features, as described below.

With this introduction, we now illustrate the methods and tool with a worked-out example using semi-notional data.

The Alternative Strategies

We consider the four alternative strategies described in Chapter Four, which are summarized briefly in Table 5.1.

The names of the strategies convey their philosophy and emphasis, but recall that all strategies are intended to be comprehensive and responsible. None are straw man extremes. After all, the United States will continue to have worldwide interests and responsibilities, none of which are likely to go away. The problems in Iraq and Afghanistan will continue for some time and, more generally, the international effects of violent Islamists will continue and perhaps worsen. This might lead to instabilities and even revolutions throughout the greater Middle East and South Asia. Al Qaeda's ambitions to mount additional attacks on the United States itself, as well as on its allies and interests abroad, are all too familiar. Iran may continue to be a source of tension and

Table 5.1
Illustrative Strategies

Strategy	Key Features
Analytic Baseline	Today's program (as of 2007), less supplementals and ground-force increases; akin to the program for the earlier 1-4-2-1 force-sizing strategy.
Direct GWOT/COIN	Emphasis: counterterrorism and counterinsurgency in the Muslim world with continued heavy intervention by U.S. ground forces likely.
Build Local, Defend Global	Emphasis: GWOT/COIN, but with local forces in the lead and U.S. ground forces primarily in a supporting role; robust investment in building partnership capacity.
Respond to Rising China	Emphasis: long-term strategic competition with China, with expectation of avoiding conflict and arms races; a strategy of "containment" of Salafism with U.S. GWOT/COIN forces largely "offshore."

threat within the Middle East, particularly to Israel; at some point it may also threaten Western Europe with nuclear missiles in the event of crisis. Looking to the east, North Korea will probably continue to be a threat to both South Korea specifically and, through its nuclear capabilities, to Japan. China and Taiwan may continue to be at odds, with the continuing potential for crisis or even conflict—albeit, a conflict that no one wants. And, inexorably, China is becoming a major military power in East Asia and beyond. Even in the near- to mid-term, U.S. forces in the Pacific are changing deployment patterns and concepts of operation for times of crisis because China's military reach is extending. Russia, although militarily weak currently except for her strategic nuclear weapons, may at some point seek to coerce or act against the Baltic states or Ukraine. Even in the Western Hemisphere, the United States has potential problems. What will happen to Cuba after the death of Fidel Castro remains a worry, the drug cartels of Latin America are a constant source of problems, and border control is looming as an increasingly important challenge.

Given such worldwide interests, responsibilities, and challenges, the question for strategy is not which to ignore (a luxury that decisionmakers do not have), but how best to posture and employ the capabilities that America has and how best to invest in future capabilities and activities. The issue, then, is a balancing act. It is for this reason that the language of portfolio management applies naturally.[10]

In the following section, we first assess the strategies for their effectiveness and risk; the next section then compares them by cost and cost-effectiveness.

---

[10] The portfolio approach to defense planning was first suggested in the mid-1990s (Davis, Gompert, and Kugler, 1996). The context was urging a shift away from the then-common practice of characterizing "strategies" in terms of force sizing alone (e.g., sizing for two major regional conflicts [MRCs]).

Effectiveness Comparisons

Figure 5.2 shows a portfolio-analysis summary comparing the nominally expected effectiveness of the four alternative strategies.[11] The strategies are in rows; the columns represent different portfolio categories in which to assess consequences of the strategies. Most of the columns are organized by operating unit (COCOM). The first six are for regional COCOMs, whereas SOCOM and STRATCOM have global responsibilities. The column marked National Command refers to the de facto command held centrally by the Secretary of Defense, supported by the Chairman of the Joint Chiefs of Staff, by JFCOM (the joint force provider), and by TRANSCOM (responsible for worldwide mobility). National Command worries about the ability to deal with issues anywhere in the world or with simultaneous crises, for example. All of these columns measure the projected health of a COCOM, relative to what the responsibilities are for the COCOM under the strategy.

Figure 5.2
Portfolio Summary, by COCOM

Measure	PACOM	CENTCOM	NORTHCOM	EUCOM	SOUTHCOM	AFRICOM	STRATCOM	SOCOM	National Command	Simultaneous war risk	Overall risk
Investment options	Detail	Detail	Detail	Detail	Detail	Detail	Detail	Detail	Detail	Detail	Detail
Analytic Baseline	O	Y	LG	LG	LG	O	O	Y	Y	O	O
Direct GWOT/COIN	O	LG	LG	LG	LG	Y	O	LG	Y	R	R
Build Local, Defend Global	Y	Y	LG	LG	LG	LG	O	LG	LG	O	Y
Respond to Rising China	G	O	LG	LG	LG	Y	G	Y	Y	Y	O

NOTES: Strategies and evaluations are notional. Color coding: red (R), orange (O), yellow (Y), light green (LG), and green (G) denote very poor, poor, marginal, good, and very good, respectively. Equivalent numerical scores are used computationally.
RAND MG703-5.2

---

[11] In this monograph illustrating methodology, we have merely estimated the effectiveness values subjectively, whereas in an application, they would be derived from a combination of analysis (including simulation-based analysis) and judgment. In such an application, there would be explicit relationships between features of the alternative programs and assessments (as in Davis, Shaver, and Beck, 2008).

The last two columns are aggregate characterizations of risk. The tiny markers in the top right-hand corners of some cells represent warnings, which we discuss below. They are used when the assessment is known to depend on fragile assumptions.

Again, we note that the assessments shown are straw men that we made subjectively without the benefit of in-depth study. They are intended to be reasonable and sufficient to illustrate methodology and provoke thought. The reader should not be overly distracted by specific assessments with which he may disagree. Such assessments might change after more in-depth work.

An examination of Figure 5.2 shows that the GWOT/COIN strategy is said to produce poor results (orange or red) for PACOM, STRATCOM, the risk associated with possible simultaneous conflicts, and "overall risk." All of the strategies have shortcomings, as indicated by the yellow, orange, and red cells. The shortcomings would be much worse had we not required that each strategy attempt to address each of the COCOM's concerns to at least some degree and that each strategy make an effort to mitigate risks. At the end of the chapter, we discuss how the strategies could be iterated to reduce some of the shortfalls.

If we now ask why the assessments in the first colored column came out as shown, we can zoom in or drill down to the next level of detail, as illustrated in Figure 5.3. The top left has a miniature version of Figure 5.2; we shall zoom on its first column, relating to PACOM; that takes us to the scorecard below and in the middle (reproduced also as Figure 5.4); zooming again on the first column of that scorecard (warfighting) takes us to a third level of detail, at the bottom right (reproduced also as Figure 5.5). The arrows indicate that the rightmost column from a lower-level scorecard is the "answer" at that level (the value of the aggregation), which is fed up to the next higher level. When using PAT "live," rather than looking at a document, zooming from the summary level is accomplished by clicking on the "Detail button" for the column of interest.

Note that Figure 5.4's assessment of PACOM results considers not just warfighting but also environment shaping and long-term

Figure 5.3
Schematic of Zooming

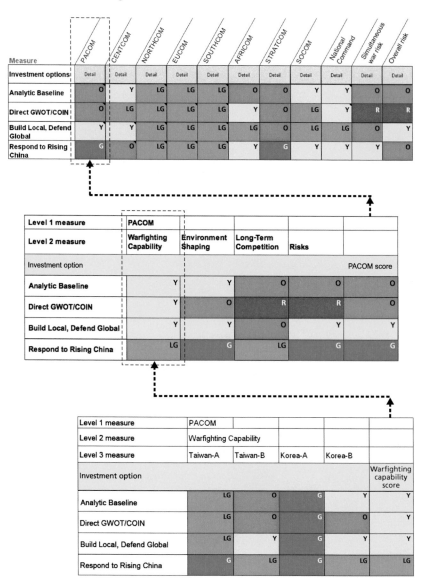

Figure 5.4
Visual Explanation of Top-Level Assessments for Asia/PACOM

Options	Factors				Assessment
Level 1 measure	PACOM				
Level 2 measure	Warfighting Capability	Environment Shaping	Long-Term Competition	Risks	
Investment option					PACOM score
Analytic Baseline	Y	Y	O	O	O
Direct GWOT/COIN	Y	O	R	R	O
Build Local, Defend Global	Y	Y	O	Y	Y
Respond to Rising China	LG	G	LG	G	G

NOTE: Strategies and evaluations are notional.
RAND MG703-5.4

Figure 5.5
Visual Explanation of Warfighting Results in Asia

Level 1 measure	PACOM				
Level 2 measure	Warfighting Capability				
Level 3 measure	Taiwan-A	Taiwan-B	Korea-A	Korea-B	
Investment option					Warfighting Capability Score
Analytic Baseline	LG	O	-G	Y	Y
Direct GWOT/COIN	LG	O	G	O	Y
Build Local, Defend Global	LG	Y	G	Y	Y
Respond to Rising China	G	LG	G	LG	LG

NOTE: Strategies are illustrative and evaluations are based on the authors' subjective judgments.
RAND MG703-5.5

competition—consistent with the evolution of strategic thinking over the last decade or so.[12]

The result in Figure 5.4 (the last column) is in this case a weighted sum of the results of the earlier columns, using relative weights of 1, 1, 1, and 2 (omitted here for simplicity). As discussed below, the aggregation rules can vary from column to column and level to level: Not only can "weights" change but so also can the rule itself. For example, the aggregation may be the worst result from the level of detail below it, or some nonlinear combination. This flexibility is important because a reviewer will sometimes want to use nonlinear rules such as: The score should be the worst of the scores of the first column, second column, and average of the other two columns.

Returning to our example, the poor result for the GWOT strategy in PACOM is due to the GWOT/COIN strategy giving relatively low priority to PACOM's region, which would mean less activity to shape the environment or compete effectively with a rising China. Obviously, the strategy could be redefined to do better in this region, but pointing out the implications of the first-cut GWOT/COIN strategy, and the need for such redefinition, is the whole purpose of this type of display. We discuss such improvements in this chapter's last section.

Zooming in on the first column of Figure 5.4 (warfighting capability) leads to the visual explanation in Figure 5.5. We see that the assessment of PACOM's warfighting is based on an aggregation of results for two China-Taiwan cases and two North Korea cases. We did not go into further detail for this illustrative work, but we had in mind that Taiwan-A would be a traditional invasion-of-Taiwan scenario and that Taiwan-B might be a more challenging test case, e.g., an invasion-of-Taiwan scenario under circumstances where U.S. forces were maldeployed and decisions delayed. Similarly, although Korea-A might be a traditional conventional invasion-of-South-Korea scenario, Korea-B might be much more complicated and might include use of

---

[12] Environment shaping was first introduced in the 1993 Regional Defense Strategy (Cheney, 1993), based on earlier RAND work (e.g., Davis, 1994b). It became a core element of defense planning in 1997 (Cohen, 1997). The same concepts appear in different words in current strategy (Rumsfeld, 2001) (i.e., the earlier elements of "assure, dissuade, deter, and defeat").

weapons of mass destruction and credible threats to U.S. bases and allies in the region. An alternative might be to consider an implosion-of-North-Korea scenario.

In principle, an infinite number of scenarios and variations might be considered, but the right way to do such work is for analysts to conduct broad exploratory analysis on the "scenario space," draw on that to identify a small *spanning set of test scenarios* that stress U.S. capabilities in all of the key dimensions, and to then evaluate results for that spanning set of test cases.[13] Here, we *assume* that a spanning set of four test cases would prove adequate (with some parameter variations within the cases). The forces of the alternative strategies could be assessed against these test cases not only by the affected COCOMs but by J-8 and PA&E using DoD's existing suite of simulations (Bexfield, 2006), and by gaming.

As previously, the higher-level result (that in Figure 5.4) can be largely understood visually by merely eyeballing results for the various factors. Note that if we had weighted the "A" cases more heavily, results would have been favorable (green). Because such relative weightings are inherently a mix of strategic judgments as well as more technical analysis, it must be easy to change these weightings easily (and it is). Further, because it is virtually certain that decisionmakers will disagree on these matters, it is important to have worked out the results for a representative range of attitudes. We accomplish this with what we call alternative *perspectives*, discussed below.

Nominal Comparison of Risks

Let us next consider risks. Different types of risk are salient in different applications (see Appendix E for a discussion of both theory and complexity). In this work, the summary assessment (Figure 5.2) has columns for two risks: the risk associated with simultaneous conflicts

---

[13] The theory and methods are described in a variety of publications (Davis, Gompert, and Kugler, 1996; Davis, 2002; Davis, Shaver, and Beck, 2008). The method was recently adopted in a National Academy study (National Academy of Sciences, forthcoming).

and a composite assessment of lower-level risks, called *overall risk*. The assessment for simultaneous conflict might be an aggregation over a number of possible simultaneous or overlapping conflicts, such as in the Middle East and East Asia, or such as a simultaneous conflict in the Middle East and a substantial attack on the U.S. homeland that strains reserve component forces, transportation capabilities, and so on. We shall not elaborate here. The column on overall risks, however, is worth further discussion. If we zoom in on the Overall risk column of Figure 5.2, we obtain Figure 5.6.

Figure 5.6 characterizes risks for all of the COCOMs' areas of responsibility. It then assesses the overall risk (last column). In this case, the aggregation is based on something more complicated than linear weighted sums. The overall risk is calculated by (1) taking the worst of the scores for PACOM and CENTCOM, (2) averaging the risks across the other COCOMs, and (3) taking the poorer score of (1) and (2). This would mean, for example, that overall risk would be rated as very high (red) if PACOM's risk were very high, if CENTCOM's risk were very high, or if the average risk across the other COCOMs were high.[14] As a result of this logic, we concluded that risks were very

Figure 5.6
Basis of Overall Risk Assessment

Option	Risk Assessment									Overall Risk
Level 1 measure	Overall risk									
Level 2 measure	PACOM	CENTCOM	STRATCOM	SOCOM	NORTHCOM	EUCOM	SOUTHCOM	AFRICOM	National Command	
Investment option										Overall risk score
Analytic Baseline	Y	Y	Y	Y	Y	Y	Y	Y	Y	Y
Direct GWOT/COIN	Y	Y	Y	Y	Y	Y	Y	Y	Y	Y
Build Local, Defend Global	Y	Y	Y	Y	Y	Y	Y	Y	Y	Y
Respond to Rising China	LG	LG	LG	LG	LG	LG	LG	LG	LG	LG

NOTE: Options and assessments are notional.
RAND *MG703-5.6*

---

[14] This illustrates why common "decision-analytic" software programs are to be viewed with suspicion. They typically assume a linear-weighted-sum logic, which leaves the user unable to do anything more subtle than change the relative weights.

high (red) for the GWOT/COIN strategy and high (orange) for the Analytic Baseline and Build Local, Defend Global strategies.

Figure 5.7 provides a visual explanation for these judgments. The assessment is an aggregation of assessments of strategic and operational adaptiveness:

1. Would the strategy in question provide adequate strategic adaptiveness for the region of interest? For example, would the relative neglect of a region mean that the United States would fail to build up the relationships and infrastructure in it to adapt quickly to alarming developments in that region should they occur? Would the strategy in question be vulnerable, e.g., to a sudden "explosion" of Salafism—the spread of radical and military Islam often characterized as in pursuit of a new Caliphate?

Figure 5.7
Basis for Risk Assessment in the Middle East (CENTCOM)

Level 1 measure	CENTCOM		
Level 2 measure	Risks		
Level 3 measure	Strategic Adaptiveness (Iran, Salafism...)	Military Adaptiveness (Disruption...)	
Investment option			Risks score
Analytic Baseline	Y	Y	Y
Direct GWOT/COIN	G	G	G
Build Local, Defend Global	LG	LG	LG
Respond to Rising China	O	O	O

NOTE: Options and assessments are notional.
RAND MG703-5.7

2. Would the strategy in question be vulnerable if events moved quickly in unanticipated ways, requiring a response with existing forces that had not been expected? That is, would the strategy allow for operational adaptiveness?

In our assessment for the Middle East, we concluded that there would *nominally* be little risk (green) for the Direct GWOT/COIN strategy because it focuses resources on the Middle East. In contrast, the Respond to Rising China strategy carries with it a relatively more sanguine attitude about the Middle East, giving it less attention and forces, which would make this strategy more vulnerable than the others to a rapid emergence of violent Islamism. Hence, the score of high risk (orange).

Uncertainty: Consequences of Different Perspectives and Assumption Sets

Assessments of a strategy's likely performance and risk depend on who is doing the assessing. Experts disagree about whether the Direct GWOT/COIN strategy would be likely to work. Some believe that it can, with enough effort and time; others believe that it is doomed to failure because the strategy's implied actions lead to what are seen by local populations as occupation by a foreign power and imposition of foreign concepts.[15] Experts also disagree on the plausibility and importance of different warfighting scenarios. As illustrated in Figure 5.8, this affects the assessment of warfighting risk. The top assessment treats CENTCOM's B cases very seriously (i.e., gives them significant weight), whereas the lower assessment does not—perhaps regarding the postulated crises or U.S. involvement in those crises as very unlikely. This illustrates how—even if detailed analysis results were precisely the same for the various test scenarios in question—risk assessment would change (compare the last columns of the two panels).

---

[15] Two recent studies discuss such matters (Gompert, Gordon, Grissom, Frelinger, Jones, Libicki, O'Connell, Stearns, and Hunter, 2008; Grissom and Ochmanek, 2008).

Figure 5.8
Alternative Assessments of Warfighting Capability in the Middle East

Nominal Assessment of CENTCOM Warfighting					
Level 1 measure	CENTCOM				
Level 2 measure	Warfighting Capability				
Level 3 measure	Iran-A	Iran-B	Oilfield Threat-A	Oilfield Threat-B	
Investment option					Warfighting Capability Score
Analytic Baseline	LG	R	G	Y	Y
Direct GWOT/COIN	LG	Y	G	G	LG
Build Local, Defend Global	LG	O	G	Y	Y
Respond to Rising China	LG	O	G	Y	Y

Assessment of CENTCOM Warfighting That Rejects Worst-Case Interventions					
Level 1 measure	CENTCOM				
Level 2 measure	Warfighting Capability				
Level 3 measure	Iran-A	Iran-B	Oilfield Threat-A	Oilfield Threat-B	
Investment option					Warfighting Capability Score
Analytic Baseline	LG	R	G	Y	LG
Direct GWOT/COIN	LG	Y	G	G	LG
Build Local, Defend Global	LG	O	G	Y	LG
Respond to Rising China	LG	O	G	Y	LG

RAND *MG703-5.8*

A Modern Approach to Cost-Effectiveness Analysis

The Dubious Concept of Cost-Effectiveness Analysis Across Categories of Goodness

One common aspect of integrated analysis is developing cost-benefit or cost-effectiveness comparisons. In many applications, this is both feasible and valuable. At a personal level, we make such comparisons when purchasing an automobile or computer; corporations make such comparisons when considering alternative organizational processes, suppliers, or partnerships; and DoD routinely conducts cost-benefit analyses when making major weapon system acquisition decisions. How to go about doing cost-benefit analysis is taught in graduate schools—although with diverse definitions of what "it" is. Unfortunately, routine cost-benefit analysis is often of dubious value and can even be counterproductive in strategic analysis.

Shortcomings of Single-Measure Effectiveness Scores. One key problem is that when the choice to be made must account for numerous factors that are different in nature (rather than different components of income or expense, for example, which can all be expressed in the same economic terms, i.e., monetized), attempting to assess the composite or "net" effectiveness with a single score is conceptually problematic. Many mathematical techniques exist for doing so, but if expressed in correct theoretical terms, the underlying problem is often complex and nonlinear. In systems engineering, this arises with *critical components*—i.e., components that must each be present and effective, independent of whether the other components are present and effective. One cannot compensate for a poor computer monitor by buying a better keyboard. In defense strategy, problems in the Asia-Pacific region will not go away merely because the nation invests more heavily in solving problems in the Middle East.

Uncertainty and the Need for Exploratory Analysis. Another deep problem is that any such assessment depends on a myriad of assumptions and judgments. For what wars should the United States be prepared? What warfighting strategies would be employed if they arose, and in pursuit of what objectives? To what extent does preparation for a given type of war help to deter or dissuade, so that the wars remain very unlikely? Such problems arise consistently in strategic-level work and cannot be resolved by merely working harder to do some calculations "correctly," using the "correct" databases and assumptions. No such things exist.

The analytic "solution" to these problems is not so much a solution as the embrace of humility and an effort to help strategic-level decisionmakers cope with complexity rather than to annoy them by offering up quantitative conclusions that obscure the very factors they must agonize about. In the present context of offering a portfolio-analysis methodology to assist the Chairman of the Joint Chiefs of Staff in making resource-informed characterizations of alternative strategies, a crucial element of the approach should be the exploratory analysis (as described above).

In the next subsection, we present a "nominal" cost-effectiveness analysis of the illustrative strategies. The following subsection then sets up a method for presenting results that do a much better job of summarizing issues for strategic-level decisionmaking, which includes a version of exploratory analysis.

A Nominal Analysis

Looking back to Figure 5.2, the summary-level scorecard, it is easy to add up the scores corresponding to the colors to produce a composite effectiveness across the categories (effectiveness of the various COCOMs, National Command, and some measures of global risk). In the case of equal weighting, a red and a green would average to a yellow. The effectiveness can be expressed as a number and divided by the cost of the strategy to obtain a cost-effectiveness ratio.

From the idealized perspective of someone teaching a course in cost-effectiveness analysis, the result might be as displayed in Figure 5.9.[16] It imagines comparing a number of options, the best of which imply a curve of effectiveness versus cost as shown. This result conveys a great deal of information: Options A and C are both desirable, at the budget levels shown, whereas option B is only modestly more effective than A and much more expensive. Options D and E are not competitive at any budget level. In the idealized case, the effectiveness and costs of the options are well known, so these conclusions about options A, B, C, D, and E are solid.

---

[16] Other idealizations exist. For example, if the options being compared have equal cost (e.g., the budget provided) or equal effectiveness (much more dubious in a world of multiple objectives), comparisons are more straightforward. In practice, the strategic options under consideration typically vary in effectiveness, cost, and the time lines on which their benefits would be achieved.

Figure 5.9
An Idealized Output of Cost-Effectiveness Analysis of Alternative
Strategies Reflecting Different Perspectives and Assumptions

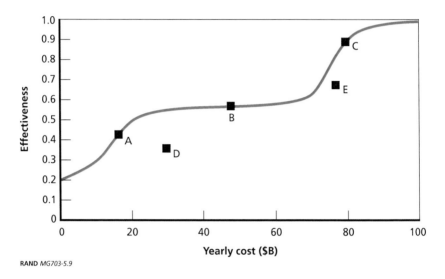

RAND MG703-5.9

In strategic analysis such as contemplated in this study, things are not so simple. Disagreements exist about effectiveness values, costs are uncertain, and even the relative goodness of the options differs depending on who is doing the evaluating. Despite all this, exploratory analysis can at least clarify the issues and, in some cases, suggest relatively robust conclusions.

Reflecting Different Perspectives and Assumptions

Given large uncertainties, the preferred approach is to employ exploratory analysis in an effort to find options (in the current case, strategies) that will achieve good results under a wide range of assumptions and preferences. This is more than traditional sensitivity analysis because the exploratory analysis does not take any particular baseline set of assumptions as a "best estimate" and actually examines the entire space of possible assumptions (albeit usually with sampling methods). This is important because, in practice (1) Baseline assumptions as may be found in official scenarios are not usually what a rational person would consider a "best estimate," but rather some bizarre mixture of best-case and worst-case assumptions, as well as deeply embedded assumptions about objectives, values, and context; (2) the uncertainties are highly correlated in the real world because, for example, an adversary who can find weaknesses will tend to exploit as many of them as possible, not merely one at a time.

The method of exploratory analysis has been developed at RAND over the last decade or so in connection with capabilities analysis,[17] but extending the theory for the purposes of portfolio analysis is challenging (Davis, Shaver, and Beck, 2008) because portfolio analysis encompasses additional types of uncertain inputs, which include

1. the relative emphasis of different objectives (e.g., warfighting, environment shaping, and development of future capabilities) and measures of option goodness
2. core subjective assumptions about the likely efficacy of complex strategies (e.g., is taming the Middle East simply a matter of U.S. expenditures and manpower?)
3. assessments about strategic, programmatic, and technical risks
4. assessments about upside potential[18]
5. performance "requirements"
6. costs.

In the current study, these issues have particular salience, as they will each time the chairman contemplates his assessments and recom-

---

[17] Some of the earliest work was done by RAND for the Joint Staff (Davis and Finch, 1993) in a study encouraging greater adaptiveness in operations plans and defense planning scenarios. A related issue paper influenced the Quadrennial Defense Reviews of both 1997 and 2001 with a discussion of RAND's approach to capabilities-based planning (Davis, Gompert, and Kugler, 1996; Davis, 2002). See also a recent study for the U.S. Navy (National Research Council, 2005). A mostly independent but parallel stream of RAND research has been applied by RAND colleagues to social-policy problems and refers variously to exploratory modeling and robust adaptive planning (Lempert, 2002; Lempert, Popper, and Bankes, 2003; Lempert, Groves, Popper, and Bankes, 2006).

[18] In this monograph, we do not include estimates for the upside potential of the various strategies because, candidly, no further upsides were evident. In an actual study, however, this aspect of the methodology should definitely be included.

mendations about alternative national military strategies to the Secretary of Defense. Ideally, exploratory analysis would consider simultaneous variations in all of the above and would identify which strategy was most robust. The theory and methods for that are not yet developed, but a first cut can be made.

A First Cut at Exploratory Analysis for Alternative Military Strategies
We suggest the following method for a first cut at exploratory analysis of alternative strategies within a portfolio framework:

- Define alternative coherent *perspectives* to distinguish among legitimate but different attitudes and assumptions that decision-makers bring to the table.

In past work we have done something like this, but we have limited the differences in perspective to be what amount to different relative weights on different measures of goodness. The analogue in the present work would be to have perspectives that differed about the weights to be placed on the needs of the various COCOMs. However, we concluded that to do so would not be enough because much more is at stake than merely "weights." For example:

- Someone inclined to emphasize one region over another may also be inclined to worry much more about worst-case scenarios in the region of primary interest to him than in other regions. Analytically, that means that people with different regional emphases will also differ about the analysis used to characterize adequacy of capability.
- Similarly, someone inclined to emphasize a particular strategy, such as a particular approach to GWOT/COIN, will likely be inclined to make optimistic assumptions about its success—if merely the resources are provided. In contrast, someone favoring a different strategy will do so in part because he believes that the other strategy is doomed to fail even with the requested resources.

- Someone inclined toward a national, global view (such as those in the Joint Staff or OSD) would be inclined to worry not only about the "balance" of effort across the COCOMs but also about the risk of simultaneous wars and the effect on deterrence if the United States were perceived not to have simultaneous-war capability. By and large, he might also be more inclined to worry about risks— region by region and globally—than would someone focused primarily on a given region or set of functional needs.
- Some people, including those who remember earlier periods in which the USAID budget was larger, would be skeptical about the "payoff function" for increased foreign and security assistance.

With these considerations in mind, we have created four *extended perspectives* as summarized in Table 5.2. The top row identifies the alternative perspectives. The first column identifies particular elements of the overall analytic structure that will be varied. The row entitled A, under CENTCOM, refers to the measure of strategy effectiveness for the class-A scenario of CENTCOM. If we read down the first numerical column, it says that the CENTCOM-leaning perspective gives relative weights of 3, 1, 0.5, 0.5, and 0 to the categories for CENT-COM, PACOM, National Command, Simultaneous wars, and Overall risks, respectively. Also in this CENTCOM-oriented perspective, assessments of the likely consequences of strategy are optimistic for CENTCOM (last row). In the second column of numerical values, the CENTCOM-leaning extended perspective gives relative weights of 1 and 3 to the A and B scenarios used for evaluations of effectiveness in CENTCOM. The numbers correspond to weights. The last row, referring to degree of optimism, is different. Without bothering the reader with details of implementation here, we have represented degree of optimism analytically by varying the assumed scores for certain effectiveness and risk values. For example, a "very skeptical" perspective will assess the GWOT/COIN strategy in the CENTCOM region as likely to produce only a marginally good outcome (yellow) and as having high risks because direct intervention can have counterproductive results. With this type of structuring, a fairly rich set of extended perspectives can be defined.

Table 5.2
Weighting Factors and Degrees of Optimism for Alternative Extended Perspectives

	Perspective			
Measure	CENTCOM Leaning	PACOM Leaning	JCS Conservative	JCS Optimistic
CENTCOM	3	1	1	1
A	1	1	1	1
B	3	0	1	0.5
PACOM	1	3	1	1
A	1	1	1	1
B	0	3	1	0.5
National Command	0.5	1	1	1
Simultaneous wars	0.5	1	1	1
A	1	1	1	1
B	0	1	1	0.5
Overall risks	0	1	1	1
Degree of optimism	Optimistic about own preferred strategy for Middle East	Very skeptical about interventionist strategy in Middle East	Cautious	Optimistic

NOTES: The numbers are weighting factors. For a given perspective, the leftmost column shows weighting factors at the top level of portfolio analysis; the next column shows weighting factors that apply at "level 3" of the analysis, where warfighting effectiveness is considered for different scenarios labeled A and B.

We performed an illustrative exploratory analysis with the results indicated in Figures 5.10, 5.11, and 5.13. Figure 5.10 shows the consequences for effectiveness of weighting the different columns of Figure 5.2 in different ways that we refer to as PACOM-leaning, CENTCOM-leaning, JCS-conservative, and JCS-optimistic. None of these perspectives represent "zealots," because real-world combatant commanders are seldom zealots; they all serve the nation, have typically done tours in different theaters or in national headquarters of the services or the

Figure 5.10
Composite Effectiveness for Different Perspectives

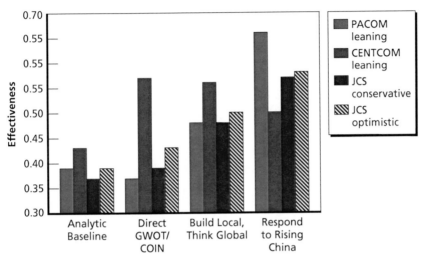

RAND *MG703-5.10*

Figure 5.11
Composite Effectiveness as a Function of Extended Perspective

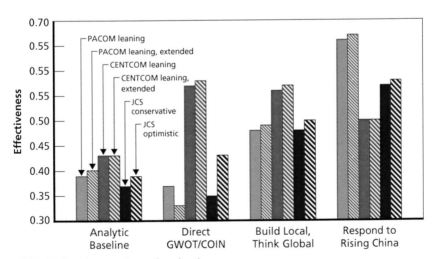

NOTE: Notional strategies and evaluations.

RAND *MG703-5.11*

Joint Staff, and are strategically savvy. Thus, they might advocate more for "their" region and they might become somewhat biased by virtue of where they sit, but they try to avoid becoming so. Nonetheless, the Chairman of the Joint Chiefs of Staff will typically be more global in his perspective and will tend to be less inclined to downplay problems in any of the regions. He may also be more sensitive to the risks associated with the possibility of simultaneous conflicts or to the importance for general deterrence of having the capability to handle simultaneous conflicts. Our two illustrative JCS perspectives differ on the margin in some of these judgments.

The differences in Figure 5.10 relate merely to weighting the different COCOMs' interests differently from one perspective to another. The evaluations of what the strategies would accomplish, however, is constant and "nominal" (in line with assumptions underlying Figure 5.2). Figure 5.11 takes the next step and shows results for both perspectives and "extended" perspectives, which evaluate the strategies differently in accordance with Table 5.2.

Two observations about Figure 5.11 are perhaps the most important. Only for the CENTCOM-leaning perspectives does the GWOT/ COIN strategy appear better than the Build Local, Defend Global strategy and, even then, not by much. For the PACOM-leaning and JCS perspectives, the Respond to Rising China strategy appears best of all. Although our evaluations are merely illustrative, they are at least plausible. And, more relevant to the methodological point, they demonstrate how the exploratory analysis can illuminate the robustness (or fragility) of conclusions.

With this background, we can now show cost-effectiveness charts for the various perspectives. Figure 5.12 does so for the extended versions of the PACOM, CENTCOM, JCS-conservative, and JCS-optimistic perspectives. It also includes the range of extraordinary operating costs that should probably be applied to the GWOT/COIN strategy. This is based on governmentwide costs, as expressed in terms of net present value of total future obligations, using a 3 percent real discount rate and an indefinite horizon. The primary observations to be made are that:

Figure 5.12
Cost-Effectiveness Comparisons with Alternative Strategic Perspectives

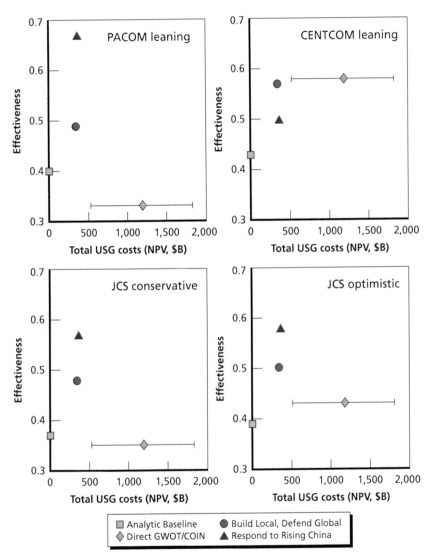

NOTES: Costs are in NPV (3 percent, indefinite horizon). Assessments are subjective.

RAND *MG703-5.12*

- Only in the CENTCOM-leaning perspective does the GWOT/ COIN strategy appear superior—even if cost is disregarded.
- More generally, the Build Local, Defend Global strategy appears superior to GWOT/COIN in cost-effectiveness because it is about as effective or more effective overall under all perspectives treated, and it saves some money even if the extraordinary costs of war/ COIN are not included for the GWOT/COIN strategy.
- The Respond to Rising China strategy is highest in effectiveness overall in all but the CENTCOM-leaning case.
- Including the noncore costs of operations (as in war or COIN), and associating those particularly with the Direct GWOT/COIN strategy, dominates consideration of cost issues.

These results, of course, are illustrative and would change if done in a "real study" based on careful analysis of fully defined strategies, programs, and cost estimates. As is typical in strategic-level analysis, it would be possible to shift results markedly with different assumptions. More important to our story, however, is the point that the strategies being evaluated here are "first-cut" strategies. After viewing results such as those in Figure 5.12, all of the strategies could be iterated so as to do better (see the next section).

To summarize the chapter so far, we have illustrated portfolio-analysis methodology to develop and present a comprehensive view of how different strategies would likely affect prospects in the different COCOMs' areas of responsibility, the risks associated with those areas, the costs, and the relative cost effectiveness. Further, we have emphasized the necessity—not merely the virtue if one has time—of employing exploratory analysis methods because of the large effects of alternative perspectives and assumption sets.

Although we have illustrated a few aspects of exploratory analysis here, additional dimensions can be quite important in some applications. Examples include exploring the uncertainty in costs and the "requirements" that may be assumed in analyses for separate theaters of operation or functional areas. Our own conclusions are as follows:

- It would be an illusion to imagine that straightforward (single-set-of-assumptions) cost-effectiveness analysis can be done on matters of higher-level strategy.
- Sophisticated decisionmakers should be shown portfolio-style results (such as those in Figure 5.2 and such drill-downs as they find useful) and the results of exploratory analysis with alternative perspectives and assumption sets. They should not be presented with single-number "net assessments" aggregating across categories (such as the COCOMs). To present them with such analysis would be a disservice.
- *After* most of the strategic decisions have been made, it may be useful to summarize and "clean up" the considerations by settling on assumptions (or test sets of assumptions) so that relationships among options can be discussed in simplified cost-effectiveness terms.

This may seem rather like heresy in some respects (i.e., suggesting that "analysis" should follow decisions), but that is not the case. The meaningful analysis, including exploratory analysis, occurs first. It is merely the "display of conclusions" that is to be tidied up and simplified later.[19]

Using Portfolio Analysis to Improve Strategies

The Need to Balance the Risks Better

The final step in our discussion of methodology is to address the question of how to construct good strategies or to improve strategies after analysis has revealed their shortcomings. Let us return to Figure 5.2, repeated here for convenience as Figure 5.13, but with a column showing total costs to the U.S. government of the

---

[19] The reader may compare this with looking at a scorecard comparison of consumer products in a magazine, noting that he does not agree with the wrapup scores, puzzling about why, recognizing his own values and assumptions, and then perhaps commenting: "Pretty good article except that it should have emphasized the first factor more than the second and third."

Figure 5.13
A Summary Portfolio Assessment

Measure	PACOM	CENTCOM	NORTHCOM	EUCOM	SOUTHCOM	AFRICOM	STRATCOM	SOCOM	National Command	Simultaneous war risk	Overall risk	USG Costs (NPV)
Investment options	Detail	Detail	Detail	Detail	Detail	Detail	Detail	Detail	Detail	Detail	Detail	
Analytic Baseline	O	Y	LG	LG	LG	O	O	Y	Y	O	O	0
Direct GWOT/COIN	O	LG	LG	LG	LG	Y	O	LG	Y	R	R	513
Build Local, Defend Global	Y	Y	LG	LG	LG	LG	O	LG	LG	O	Y	342
Respond to Rising China	G	O	LG	LG	LG	Y	G	Y	Y	Y	O	372

NOTE: Costs for Direct GWOT/COIN do not include the extraordinary costs discussed in Figure 5.12.

four strategies (expressed as the net present value of future obligations, using a 3 percent discount rate after inflation). On viewing this result, a senior leader such as the Chairman of the Joint Chiefs of Staff could note the large additional price tag of the GWOT/COIN strategy and ask whether it could be reduced to be comparable to the others. Alternatively, and more likely, he would first be concerned about whether any of the strategies could be "fixed" by addressing the problems that show up as reds and oranges, and perhaps even yellows. What could be done by starting with the GWOT/COIN strategy but adding to the hypothesized actions some additional initiatives to reduce problems in PACOM and those posed by the possibility of simultaneous war? Similarly, is there a variant to the Respond to Rising China strategy that would do better in CENTCOM? Is there a variant to the Build Local, Defend Global strategy that would address STRATCOM's most serious needs and reduce the risk factor relating to simultaneous wars?[20] When constructing our original illustrative strategies, we included items attempting to address problems in COCOMs other than the one focused on in a given strategy. Had we not done so, the equivalent of

---

[20] PAT does not itself answer any of these questions. Rather, it is a tool in which analysts can embed the results of work on the adequacy of a strategy's forces for warfighting, shaping, and so on. Some of that work might be accomplished in DoD's "availability studies"or similar efforts; other work would need to be based on structured expert judgment.

Figure 5.11 would have reds in numerous cells. It appears that we did not go far enough, however. And, in fact, there are ways to improve the various strategies—usually with a price tag but not always as much of one as might be thought.

Hedges to Improve Strategic and Operational Adaptiveness

The classic theory on how to go about such matters is to consider equal-cost alternatives. Thus, one might consider budgets of, say $200B and $550B in NPV terms, and ask that the developers of the alternatives see how they could make their strategies better—in terms of this global view—for each of those budget levels. *All* sensible strategies are "hybrid strategies" in that they address all of the COCOM issues to at least some degree, and with iteration the strategies would all take on more of a hybrid character, although they would still differ in important respects.

Someone attempting to improve a given strategy in response to such a global-minded request would look first to mitigate serious problems with inexpensive investments or relatively painless shifts of existing resources. He might also try to influence how the evaluations are done, such as by pointing out that a COCOM's reported problems are due to what may be an overemphasis on a stressful warfighting case. In addition, he would seek to mitigate risks by providing for hedges to improve strategic and operational adaptiveness. We included some such hedges even in our illustrative first-cut strategies. They include, for example:

- Maintaining operational reserves, within each COCOM and nationally, rather than reducing force structure excessively because of a best-estimate belief that less force structure would be adequate. It could reasonably be argued that we did not go far enough, and that the final version of the options should not reduce existing force structure at all.[21]

---

[21] The primary point of disagreement relates to ground forces. Some argue that with Saddam Hussein toppled, Europe stable, and South Korea able to defend itself on the ground, there is less need for U.S. ground forces except in the event of "more Iraqs," which should be studiously avoided. Others are more conservative, believing that future intervention in

- Avoiding vacuums, by maintaining enough force presence and sufficiently great levels of engagement so that potential trouble-makers do not smell opportunities.
- Maintaining vigorous research and development, and infrastructure, to permit larger-scale strategic adaptations if those should be necessary (an adaptation, for example, to a global "explosion" of Salafism on the one hand or to more aggressive military build-ups by China). In some cases, R&D (plus appropriately visible testing) might be sufficient to suppress the emergence of threats that would otherwise require expensive acquisition of additional force structure or weapon systems.

Although we have merely brushed the surface with our discussion, we hope that it is adequate to convey the central ideas. Let us conclude this chapter on a philosophical note.

Summary Objective: Strategy That Is Flexible, Adaptive, and Robust (a "FAR Strategy")

The fundamental purpose of reviewing and iterating strategies as we have described is to identify those that are as flexible, adaptive, and robust (FAR) as possible. Such FAR strategies are far superior to strategies imagined to be "optimal" but that are in fact sensitive to all kinds of dubious assumptions.[22] The terminology of FAR strategy is chosen carefully: "Flexible" refers to the ability to undertake missions above and beyond those focused on during planning; "adaptive" refers to the ability to be effective in a very wide range of operational circumstances

---

manpower-intensive wars may be necessary, whether or not strategy in 2008 is premised on avoiding it. In their view, the current ground-force structure is "about right" in total size even if its nature should be changed. There are disagreements even among the present authors on such matters.

[22] This emphasis traces back to a much earlier study for the Joint Staff (Davis and Finch, 1993). It is a consistent emphasis in a body of recent work, including National Academy work advising DoD on its approach to modeling, simulation, and analysis (National Research Council, 2006).

(i.e., different scenarios, with different contexts and objectives, as well as facts on the ground); and "robust" refers to the ability to withstand and recover from adverse shocks. Although all of these attributes are sometimes lumped together by referring to adaptiveness, robustness, agility, or a number of other words, we suggest distinguishing among them because the differences are indeed significant.[23]

Is the emphasis on FARness a mere expression of approval for motherhood? Hardly. Until about 1980–1983, U.S. force planning focused so much on NATO's Central Region as to leave a vacuum of capabilities in the Persian Gulf. That shortcoming was remedied with creation of the Rapid Deployment Joint Task Force and, later, CENT-COM, which proved crucial in 1990–1991. In the late 1990s, there was enthusiasm in some quarters for cutting the size of the Army further because manpower-intensive wars were regarded as very unlikely—an assumption that proved false after the attack of September 11, 2001. The operational planning for Operation Iraqi Freedom was fatally flawed by its failure to take seriously the possibility of difficulties in the stabilization phase. In going about global force planning, DoD needs to be humble about its ability to predict the nature or location of future crises and conflicts. This plea for humility can be overdone in that the list of plausible adversaries is reasonably limited and the geostrategic environment does not actually change all that rapidly, but because the continuing tendency is to take standard planning scenarios much too seriously, the admonition for FARness is important.

---

[23] For the one-word summary, we have used "adaptiveness" (Davis, Gompert, and Kugler, 1996); some RAND colleagues refer to "robust adaptive planning"; and OSD's David Alberts has often used the term "agility." To a good approximation, all of these have the same ideas in mind.

CHAPTER SIX

Conclusions

Methodology, Tools, and Analysis

Previous chapters have sketched and illustrated the methodology that we have developed for an integrated characterization of alternative strategies' most likely effectiveness, risks, and resource implications. The strategies, assessments, and costs used were illustrative—drawing on issues and options that are very much uncertain as of late 2008, as well as considerable approximate information on costs.

We believe that the methodology is sound, integrated, and sufficiently simple to be used in studies in support of the Chairman of the Joint Chiefs of Staff, the Secretary of Defense, and other high officials. We believe that it is well suited to providing a basis for the chairman's resource-informed recommendations to the Secretary of Defense and the President about national strategy. Ultimately, however, that will be for others to judge.

The methodology is relatively straightforward superficially, but its value can be thoroughly undercut if applied mechanically or with a premium on consensus and best-estimate judgments. Much of the strength of good strategic planning depends on highlighting precisely what people find uncomfortable and what staff analysts often assume would be unacceptable to senior leaders. The purpose should not be to convey a feel-good sense of the various strategies and their prospects, or to reinforce biases, but rather to convey a sense of the strategies' strengths and shortcomings and to assist in refining them so that the iterated strategies are responsible and have the potential for success. The following paragraphs describe more systematically what is needed.

Using the Methods and Tools

A natural question is how easily the methodology we have illustrated could be applied, either within or for the Joint Staff. We are optimistic, but the special nature of strategic planning suggests the need to reiterate some principles and then post some cautions.

Attributes of Good Strategic Analysis

As reflected in the observations of thoughtful analysts over the decades, a number of attributes characterize good analysis.[1] For the purposes of strategic analysis, they are arguably critical:

- *Comprehensibility and Transparency*: Strategic analysis is of little or no value unless it is truly understandable to senior leaders. That requires the ability to understand the whole and, to a significant degree, the transparency allowing assumptions and their vulnerabilities to be identified.

- *Taking an Integrative, System Perspective:* A hallmark of both systems analysis and its softer and more strategic descendent, policy analysis, is that they reflect a broad and encompassing view, rather than one dealing only with the numerous "piece parts." Something may be comprehensive, but not integrated; something may be comprehensible, but misleading for failure to address the system issues effectively. For strategic analysis in the Joint Staff, close cooperation will be needed between the J-8 and J-5 directorates in particular, and among others on specific issues.

- *Objectivity*: This is an aspiration rather than something that one can achieve to perfection, but it can be approached and the value of doing so is enormous in strategic planning. It can be undercut

---

[1]   It is interesting to compare notes on such matters with classic sources on systems and policy analysis (Kahn and Mann, 1957; Quade and Boucher, 1968; Quade and Carter, 1989; Morgan and Henrion, 1990).

by parochialism, conflicts of interest, or lack of suitable subject-area knowledge or analytical depth.

- *Candor*: Often not mentioned explicitly, it is critical in strategic planning because the stakes are so high and because the toughest issues and most difficult choices are all too easy to leave undiscussed because of controversy or discomfort.
- *Interactiveness*: Strategic analysis is not suited for being "handed over the transom" but rather is something to be developed, presented, discussed with, and iterated with senior leaders. Strategic planning is typically a learning process—one that uncovers values, criteria, and new options as the process continues.
- *Reliability*: The analysis should be "sound" to the extent feasible in the time available; further, part of analysis is to assess its own reliability.
- *Treatment of Uncertainty:* This issue has long been more poorly treated than other issues when bringing to bear analytic methods. The problems in dealing well with uncertainty have been noted for decades. Modern developments provide numerous powerful methods for dealing well with uncertainty (Morgan and Henrion, 1990; Davis, 2002; National Research Council, 2005), but they are often resisted by senior leaders, analytic organizations who find them difficult to apply with their familiar tools and methods, or both.

Challenges

The challenges of meeting all the attributes of good analysis are many and largely well known, but we mention some in particular because they bear on whether the methods and tools we have presented are usable.

Committees or Small Groups? As the number of participants or reviewers of analysis increases, the time required can explode and the value of the results can plummet. The committee approach to analysis has severe shortcomings for strategic analysis, including ill effects on candor.

Review and Concurrence. The history of good analysis organizations as we understand it shows that a key challenge is to have the work done by a small, first-rate group that is protected and empowered, followed by extensive review, followed by revisions, with final decisions made by the analytic group. Unless it "owns the typewriter," to use an ancient description, the result of the broad review will be to water down the study on precisely the items on which the most candor and clarity are needed. It is better to pass along negative reviews and disagreements than to dilute analysis by accommodating to achieve concurrence among battling factions.

Timeliness Versus Depth. Studies can be done very fast (in days or weeks), fast (in a few months), or slowly, with some obvious tradeoffs. The methodology we have described can be applied on any of those time scales given background knowledge and data. However, the faster the analysis needs to be done, the more essential it is to approach evaluations with expert judgment at relatively high levels of aggregation. That approach, which we used in this study that was merely illustrating the methodology, is sufficiently fast and simple that it can be understood and reviewed. Any effort to incorporate layers of detail, however, particularly levels of detail based on complex models, will carry the risk of deeply buried error, as well as the likelihood of confusion and difficult-to-explain results. Given a longer period of time, however, it should be quite feasible to draw on large ongoing analytic activities, such as appear in the Department of Defense's Analytic Agenda to improve the quality and depth of analysis (Bexfield, 2006). In our view of capabilities-based planning (Davis, 2002), the analytic teams would first do exploratory analysis with low-resolution models and gaming, drawing on the results to define a good test set of cases to be carried out in more detail with the best DoD models and analytic teams for the job.[2] The results of those test cases would be reflected in the portfolio analysis where, in our illustrative work, we showed outcomes for A and B versions of warfights in different COCOMs' areas of responsibility.

---

[2] This may seem the same as current practice, but DoD's official planning scenarios have not typically been developed with an eye toward their constituting an analytically appropriate spanning set in the sense that we use that term (Davis, Shaver, and Beck, 2008).

Next Steps for Research and Applications

The next steps in pursuit of the approach we have described should include (1) refining the analytical tools (PAT and a related tool called BCOT (Davis, Shaver, Gvineria, and Beck, 2007) so that they can be more easily used by Pentagon and COCOM staff analysts, and (2) using the methodology in an applied study with more carefully defined strategies and more in-depth analysis. Such a study should, in particular:

1. develop a wider and more carefully articulated set of strategies to reflect the full range of views likely to be expressed in the upcoming national debate
2. elaborate on the concepts of preparing for strategic and operational adaptiveness and translate those into related programs, force shifts, and other initiatives
3. develop methods for evaluating effectiveness and risk that would draw either on analytical models (including those associated with DoD's Analytic Agenda) or on the structured judgment of experts when they are presented with enough contextual information to make their judgments meaningful
4. develop well-defined spanning sets of "cases" (scenarios with particular assumption sets) to test alternative strategies in all of the important dimensions
5. define a less-comprehensive Analytic Baseline strategy so that more aspects of the current DoD program would be "on the table" for strategic-level tradeoffs.

The first two of these would be in the natural province of the Joint Staff's J-5 and OSD's Policy office. The others would be of particular interest to the Joint Staff's J-8 and to OSD's office of Program Analysis and Evaluation. To the extent that some of the strategies and potential adaptations would depend on new military capabilities, the work would relate closely to the work of OSD's Acquisition, Technology, and Logistics (AT&L).

Responsibilities of the Chairman of the Joint Chiefs of Staff

The Chairman of the Joint Chiefs of Staff is the principal military advisor to the President, the National Security Council, and the Secretary of Defense and is also the spokesman for the combatant commanders, especially with regard to their operational requirements. Among his many responsibilities, he is expected to

- integrate information from the combatant commanders on their military requirements, separately and in total, and recommend priorities to the Secretary of Defense regarding those requirements and priorities
- advise the Secretary of Defense on how well the program and budget proposals of the military departments and other DoD components address the expressed requirements
- submit alternatives to the Secretary of Defense—alternative proposals that work within the secretary's guidance (including fiscal guidance) to better address the prioritized requirements
- assess military requirements for DoD acquisition programs.

These paraphrased responsibilities are summarized in a recent Instruction on Capabilities Based Planning (Chairman of the Joint Chiefs of Staff, 2007) and are based on public law.[1]

---

[1] U.S. Code, Title 10, Sections 151, 153, and 163.

Aspects of the Congressional language and formal instructions must be interpreted. Read naively and in isolation, some of the language would suggest that "requirements" are whatever commanders, military departments, and defense agencies say they are. In practice, the Secretary of Defense regards the statement of requirements as requests—requests based on responsible assumptions to be taken quite seriously, but assumptions that must be questioned. After hearing advice from others, the secretary may conclude that a given combatant commander needs more or fewer resources or that the requests are sound but in excess of what the President and Congress are willing to spend. Or he may conclude that adjustments are needed because of impending changes in national security strategy and national military strategy. It such cases it is then necessary to prioritize efforts to address the requirements, persuade requesters to change their requirements, add additional features to the program, or some combination. The process is inherently iterative with multiple voices and considerations. As the principal military adviser, the Chairman of the Joint Chiefs of Staff plays a central and crucial role in sorting matters out and in advising on resource-informed strategic decisions.

The Joint Staff's Directorate for Force Structure, Resources, and Assessments (J-8), in turn, is the primary organization for assisting the chairman as he develops resource-informed assessments and recommendations on many of the core issues.

The Portfolio Analysis Tool

Origins in Concepts for Post–Cold War Defense Planning

This appendix provides more background on RAND's portfolio analysis tool (PAT). PAT's origins lie in RAND research performed over many years to improve theory and methods for defense planning.

In the early 1990s, RAND researchers labored to develop concepts and methods for post–Cold War defense planning. The intent was to confront forthrightly the multifaceted nature of the challenges to be faced and also the ubiquitous role of uncertainty. This was in contrast to having strategy and planning focus on preparing for one or two stereotyped warfighting scenarios. The evolving methods were described with different names, such as uncertainty-sensitive planning (Davis, 1994d), planning for adaptiveness (Davis, Gompert, and Kugler, 1996) and capabilities-based planning (Davis, 1994a). An important element was the suggestion that defense planning should have a portfolio-management framework to help decisionmakers see readily across the full range of DoD's responsibilities when reviewing alternatives (Davis, Gompert, and Kugler, 1996). The suggested framework in 1996 encouraged characterizing alternative defense strategies in terms of these abilities:

- *environment shaping*
- providing *capabilities* suitable for deterrence, crisis response, and warfighting in a diversity of scenarios and circumstances

- assuring *strategic adaptiveness* over time through suitable research and development, infrastructure development, and other preparatory actions.

In this framework, preparing to fight two simultaneous wars was an important part of the second objective, but definitely not the exclusive focus.

The concepts of both capabilities-based planning and related portfolio analysis were further refined over time and were summarized in a 2002 monograph (Davis, 2002).

As discussed in the main text, three successive administrations have drawn on the principal ideas, each in its own way and with each adding its own ideas and drawing also on numerous sources. The concept of environment shaping was used in the Regional Defense Strategy of the first Bush administration (Cheney, 1993). The Clinton administration's first Quadrennial Defense Review adopted what amounted to the recommended portfolio-management perspective in a strategy called Shape, Respond, and Prepare Now (Cohen, 1997), and similar features are incorporated in the current Bush administration's strategy of Assure, Dissuade, Deter, and Defeat. The Bush administration fully embraced capabilities-based planning (Rumsfeld, 2001) with its emphasis on planning under uncertainty. DoD now sees much of its planning in portfolio terms and has referred to it that way in statements by the secretary, senior officials, and the Vice Chairman of the Joint Staff. Suggestions for different ways to apply portfolio management come from sources as diverse as the Defense Science Board (Defense Science Board, 2007) and the Government Accountability Office (Government Accountability Office, 2007), although the suggestions often are unclear

Tools for Portfolio Analysis

Importance of Tools

Concepts and theory are fine, but practical analysis and planning depend significantly—for good and for bad—on the tools used. These can facilitate work and shape its character. With this in mind, RAND researchers have for some years been developing tools for portfolio analysis. An early tool was DynaRank (Hillestad and Davis, 1998), developed in the mid-1990s during a project proposing and comparing alternative defense strategies in anticipation of the first Quadrennial Defense Review (Davis, Kugler, and Hillestad, 1997). DynaRank has subsequently been improved through a number of defense and social-policy applications. In 2005, a subsequent tool (PAT-MD [missile defense]) was developed for strategic planning in the Ballistic Missile Defense Agency (BMDA) (Dreyer and Davis, 2005). After seeing a presentation on work for the BMDA, the Under Secretary of Defense for Acquisition, Technology, and Logistics asked RAND to develop a generic version. The result was RAND's Portfolio Analysis Tool (PAT) (Davis, Shaver, and Beck, 2008), which was further enhanced in the current project. PAT continues to evolve and is now being used in a study for the U.S. Air Force's program planning.

PAT's Intended Purpose

PAT assists in top-down portfolio analysis and support of decision-making. Many applications are possible, but the common motivation for using PAT is the need to characterize the relative goodness and shortcomings, cost, and cost effectiveness of alternatives—i.e., different courses of action, programs, or investment packages—intended to contribute value in a number of different categories, such as geographic theaters, warfare domains, capability areas, or such strategic categories as warfighting, environment shaping, and laying the basis for future large-scale adaptations. PAT itself is "an empty vessel," a valuable spreadsheet tool with numerous features to assist in analysis, but it depends entirely on the user's structuring the problem of interest and providing the necessary evaluation information, whether empirical or model-based. PAT then assists in laying out information accordingly. Options appear in rows, various measures of goodness appear in columns, and the various categories of the portfolio are represented by groups of columns.

PAT's Status and Documentation

PAT is an evolving tool for working analysts, rather than a glossy, polished commercial product. It is quite usable now, but it should be considered akin to "late-beta software." The original version was formally documented (Dreyer and Davis, 2005) and most of the important enhancements are summarized in the appendix of a recent study (Davis, Shaver, and Beck, 2008). The changes include an application-independent structure, a "multiresolution feature" making it possible to enter assumptions at alternative levels of detail, a much-improved ability to define and store alternative perspectives as defined in Chapter Five, and a much-improved user interface. User-manual-level documentation is no longer up to date, but this is not a serious problem in practice, assuming some communication with the chief developer (Dreyer) or another RAND user. PAT runs under Microsoft Windows XP, the Macintosh OS X system, or a virtualization program such as Parallels® running on a Macintosh. It is built within Microsoft Excel 2003, which is compatible with Microsoft Excel 2004.

A Tool for Customized Reporting

The costs for the strategies discussed in this monograph were developed using Microsoft Excel. Excel was used to take the data discussed in Appendix D—principally the 20-year cost of a given program, but also such information as when that program might take effect—and spread those costs in a plausible manner over time from 2009 to 2028 and across three classes of expenditure: research and development (R&D), acquisition, and operations and support (O&S). The information can then be readily manipulated in useful ways, such as to find the net present value (actually, a cost) of future obligations, or to apply a burdening factor that accounts for personnel-related costs that are borne outside DoD. Perhaps most useful, however, is Excel's "pivot table" feature, which generates cost reports in a variety of structures.

Excel's pivot table feature summarizes large tables of data and provides a way to readily arrange and rearrange those data. The cost data for the alternative strategies are collected in what amounts to a large table. Each strategy has between 12 and 53 programs or force shifts.[1] Each program is described by numerous categories of information. We developed six categories of cost information and five key categories of identifying information. Others could be added readily.

The types of cost information are

---

[1] When forces are moved from one COCOM to another (e.g., if 12 capital ships are moved from EUCOM to PACOM), it is useful to list this as two separate programs. The first cuts forces from one COCOM, and the second program adds those same forces to another COCOM. The overall cost implication is zero, but by listing the program once as a cut and once as an addition, it is possible to see the theater-level cost implications.

1. cost, 2009–2028 (FY 2009$)
2. FYDP cost, 2009–2014 (FY 2009$)
3. NPV of cost (3 percent discount rate)
4. NPV of FYDP cost, 2009–2014 (3 percent discount rate)
5. cost, 2009–2028 (FY 2009$), with a variable burdening factor to account for additional personnel-related costs
6. notional 20-year cost (FY 2009$).

The categories of identifying information are

1. COCOM
2. service
3. strategy (Direct GWOT/COIN; Build Local, Defend Global; or Respond to Rising China)
4. DoD or other USG program
5. unit type (e.g., Army BCT or medium-range bomber wing).

This information is arrayed in columns in a single table containing all the programs for all the strategies. Although the implications of the data are unintelligible in this state, which stretches to well over 1,000 cells of information, the table provides the raw data for pivot table manipulations. A pivot table can sort and summarize any of the cost data by any of the established categories. Multiple categories can be used to filter the data at the same time. For example, the pivot table function can generate not only the 2009–2028 (FY 2009$) cost for each strategy, but it can also show the distribution of that cost by COCOM within the strategies. The cost by COCOM can be further broken down by service. It is possible to display as many sorts of data side by side as the user would like.

Many useful pivot tables could be built to compare the costs of the alternative strategies. The five variations offered here stand out as especially illuminating:

1. a comparison of DoD and other USG costs, 2009–2028 (FY 2009$), by strategy

2. a comparison of 2009–2028 (FY 2009$) costs, by strategy and COCOM (all USG costs considered)
3. a comparison of 2009–2028 (FY 2009$) and NPV, by strategy and COCOM (all USG costs considered)
4. a comparison of 2009–2028 (FY 2009$) costs, by COCOM and service for a given strategy (only DOD costs considered)
5. a comparison of 2009–2028 costs by strategy and service (all USG costs considered).

These comparisons are provided in the tables below.

Table C.1
Cost, 2009–2028 (FY 2009$), by Strategy and for DoD Versus USG

Strategy	USG or DoD	Cost (FY 2009 $B)
Direct GWOT/COIN		**301.6**
	DoD	248.4
	USG	53.1
Build Local, Defend Global		**219.2**
	DoD	–27.6
	USG	247.0
Respond to Rising China		**253.7**
	DoD	191.0
	USG	62.7

NOTE: Numbers may not add to totals because of rounding.

Table C.2
Cost, 2009–2028 (FY 2009$), by Strategy and COCOM (All USG)

COCOM	Strategy and Cost (FY 2009 $B)		
	Direct GWOT/COIN	Build Local, Defend Global	Respond to Rising China
AFRICOM	10.5	70.6	15.4
CENTCOM	283.8	79.2	−31.8
EUCOM	—	−41.7	−90.7
National Command	—	88.4	75.1
PACOM	7.2	−11.1	155.9
SOUTHCOM	—	13.5	3.0
STRATCOM	—	20.3	126.7
Total	**301.6**	**219.2**	**253.7**

NOTE: Numbers may not add to totals because of rounding.

Table C.3
Cost, 2009–2028 (FY 2009$), and NPV of Cost 2009–2028, by Strategy and COCOM (All USG)

Strategy	COCOM	Cost (FY 2009 $B)	NPV, 3% Discount Rate
Direct GWOT/COIN		301.6	519.6
	AFRICOM	10.5	17.5
	CENTCOM	283.8	490.0
	PACOM	7.2	12.1
Build Local, Defend Global		219.2	340.6
	AFRICOM	70.6	118.5
	CENTCOM	79.2	117.8
	EUCOM	−41.7	−81.9
	National Command	88.4	168.0
	PACOM	−11.1	−36.7
	SOUTHCOM	13.5	22.8
	STRATCOM	20.3	32.1
Respond to Rising China		253.7	366.1
	AFRICOM	15.4	25.4
	CENTCOM	−31.8	−67.2
	EUCOM	−90.7	−214.9
	National Command	75.1	142.5
	PACOM	155.9	263.9
	SOUTHCOM	3.0	4.7
	STRATCOM	126.7	211.8

NOTE: Numbers may not add to totals because of rounding.

Table C.4
Respond to Rising China Cost, 2009–2028
(FY 2009$), by COCOM and Service (DoD Only)

COCOM	Service	Cost (FY 2009 $B)
AFRICOM		**3.1**
	Navy	3.1
CENTCOM		**71.7**
	Army	−75.0
	Navy	3.3
EUCOM		**−90.7**
	Army	−69.3
	Navy	−21.4
National Command		**75.1**
	Army	69.3
	Navy	5.8
PACOM		**145.5**
	Navy	83.4
	USAF	62.1
SOUTHCOM		**3.0**
	Navy	3.0
STRATCOM		**126.7**
	Navy	18.4
	USAF	108.3
Strategy total		**191.0**

Table C.5
Cost, 2009–2028 (FY 2009$), by Strategy and Service

Strategy	Army	Marine Corps	Navy	USAF	Other USG	Total (FY 2009 $B)
Direct GWOT/COIN	175.0	73.4	—	—	53.1	**301.6**
Build Local Defend Global	−83.0	—	29.3	25.9	247.0	**219.2**
Respond to Rising China	−75.0	—	95.6	170.4	62.7	**253.7**

NOTE: Numbers may not add to totals because of rounding.

Documentation of Cost Estimates

This appendix documents the programs and costs used in the main text as vehicles to illustrate the methodology. They include numerous approximations and do not reflect official Department of Defense force or cost data. However, where possible they are derived from responsible open-source analyses such as those from the Congressional Budget Office or the Congressional Research Service.

Populating the portfolio analysis tool (PAT) with representative cost numbers and finding a consistent way to carry those costs forward over time was an important task. For the strategies to have relevance in illustrating the methodology, the resources attributed to them had to be fairly realistic. PAT generates useful cost-effectiveness comparisons and shows the cost implications of the various strategies over time to the extent that it is populated with cost data that accurately illustrate the strategies. Relatively modest changes in some investment items' costs can, carried over a 20-year period, have a measurable effect on the resource implications of a strategy.

The data and assumptions that were used to project the costs of the programs that underpin the strategies are described below. First, a number of important general cost issues are addressed and the process used to arrive at the estimates is discussed.

General Issues

Some assumptions and general issues deserve special attention. First, the illustrative nature of the strategies has important implications for the cost figures. The investment decisions made within the strategies characterize the strategies' respective approaches, strengths, and weaknesses. Each strategy addresses the nation's various security challenges in a responsible way but glosses over inevitable bureaucratic, programmatic, and political constraints. This leads to total costs and cost distributions that would be challenging to execute. The overall 20-year price tag for the entire U.S. government for each strategy represents a significant increase in spending over the Analytic Baseline strategy, at a time when defense spending has reached its highest inflation-corrected level since World War II. Further, each service department has been receiving about 30 percent of the defense budget during the past decade. The strategies developed here entail significant resource gains for some services and losses for others, which would be contested.

Second, programs were assembled into units that were convenient to cost and that could serve as readily understandable shorthand for a type and scale of a given capability. For instance, unit names such as "squadron" refer to roughly similar but varying numbers of aircraft, depending on the aircraft type and the specific assumptions used to arrive at a cost figure. Moving a squadron of C-17s from PACOM to EUCOM signifies the transfer of a general airlift capability commensurate with about that number of aircraft. The requirements are not derived from detailed modeling and simulation.

Last, the cost data discussed in this section are only one part of the overall "resource implication" picture. Nonmonetary resources are vital to consider when weighing strategies against one another. Certain programs have greater nonmonetary resource implications than others. The Respond to Rising China strategy, for instance, introduces 79 new ships to the Navy, above those already programmed in the Analytic Baseline. These ships are introduced over a seven-year period between 2016 and 2023. Although U.S. shipyards have the theoretical capacity to produce roughly 30 ships a year (O'Rourke, 2004, p. 25), and many of these 79 ships are relatively small and inexpensive, in recent years the Navy has purchased only about a sixth of that total. Executing this plan would place a huge demand on the domestic shipbuilding industry and would require that new personnel be recruited and trained. The Respond to Rising China strategy also has the Air Force purchasing two new types of bombers in the same time frame, which itself would be a challenge for the defense industry (and which would complicate the serious bureaucratic, political, and fiscal challenges already raised by its sister service's large capital investments). In reviewing the 20-year cost of the programs enumerated below, therefore, the reader should also consider the implied nonmonetary resource implications.

Cost Methodology

The same basic methodology is applied to each program to develop the costs for 2009–2028. First, the notional 20-year cost was established by adding together the costs of research and development (R&D), acquisition, and 20 years of operations and support (O&S)—essentially, all life-cycle costs other than disposal. R&D costs were then ignored in those cases where the program in question was already in the force today or was part of the baseline defense plan. R&D costs in those cases will be borne by the Defense Department regardless of whether such programs are included in the alternative strategies. Acquisition costs were also ignored in cases where the units in question were already in the force. That acquisition money has been spent, and DoD would neither save money by shedding those items nor incur additional acquisition costs by moving them from one COCOM to another. As a result, the 20-year costs for a number of illustrative programs reflect only O&S.

With notional 20-year costs in hand, we estimated the shares of those costs that would go to R&D, acquisition, and O&S. This was done by expressing the information gathered (data on actual or estimated R&D, acquisition, and O&S costs) as percentages of their sum. (When existing programs did not include R&D or acquisition costs, their O&S costs were set at 100 percent of the 20-year cost.) This extra step—summing and then re-dividing the R&D, acquisition, and O&S costs—resulted in less precision, but it made the respective shares readily comparable across programs and eased the process of determining the costs over time for the numerous items.

After establishing the notional 20-year costs and how those costs would be divided between R&D, acquisition, and O&S, the costs were spread over the 2009–2028 period. We made judgments on how quickly existing resources could be shifted from one COCOM to another, on when future programs could reasonably be expected to appear in the force, and on how long a program would take to develop, procure, and deploy. This enabled the cost of a strategy to be shown over time, by type of cost.

FY 2009 was chosen as the starting point for new strategies. Hence, 2009 is the earliest year that R&D, acquisition, and (for existing units) O&S can begin. R&D and acquisition stretch for variable terms (from one to ten years) depending on the program in question. In most cases, it was estimated that if a strategy were embarked on in 2009, it would take several years to realign resources or for new units to achieve initial operational capability (IOC). With a few exceptions, O&S costs do not start until 2011.

The timing of decisions for the individual programs are discussed below, but two general points are worth making now. Most important, the actual cost of a strategy in the time period 2009–2028 is not the same as what we call the "notional 20-year cost." The preponderance of resource shifts do not begin in 2009, and those bearing on future technologies (e.g., long- or medium-range bombers) do not begin until quite late in the 20-year time frame. A considerable portion of the notional 20-year O&S costs stretch beyond 2009–2028. The "notional 20-year cost," then, is a useful metric, but it does not correspond neatly to actual spending in the 2009–2028 time period.

Second, the timing of the introduction of new weapons systems was considered individually, with the costs of those programs introduced into the strategies as soon as was reasonable. The sooner costs are shown in the 2009–2028 time frame, the less the costs extend beyond 2028 and are therefore hidden from anyone reviewing the strategy. In some cases, favoring the assumption of early introduction probably made these strategies more difficult to execute. To take the example discussed above, the Respond to Rising China strategy calls for a large number of new ships, all of which are introduced between 2016 (when R&D could conceivably be complete) and 2023. A less-aggressive procurement schedule would be easier to accomplish but, in pushing toward or beyond the 2028 horizon, such a schedule would have obscured a meaningful part of the total cost of the strategy. In such cases, we chose to show costs as soon as seemed feasible. An alternative would have been to show cost streams out well beyond 20 years and to use net present value calculations. That would have had both advantages and disadvantages—a major disadvantage being that few readers of this monograph are likely to "think," or to have data for more than, 20 years into the future, or to talk in NPV terms. In the main text, we do show net present values of the future obligations of strategy. Because of having assumed earliest-reasonable introduction of forces, the NPV figures are overestimated in some cases.

Our calculations also captured a rough approximation of the recapitalization costs of each strategy. Although no program-by-program research was done on recapitalization, it was important to express the recurring costs that would be associated with a strategy's capital investment. None of these recapitalization costs appear in the 2009–2028 time frame. However, they all increase the cost of the strategies when the full NPV of the costs is given.

Recapitalization costs were established by distributing the original acquisition cost for each program over the out-years. This distribution was determined by the general frequency with which a program might expect to be recapitalized: The acquisition costs of ground forces were spread over 20 years and the costs for air and naval forces were spread over 30 years. The start of recapitalization was sensitive to the year in which the program was initially acquired. Many programs begin to recapitalize in or around 2028, but those that were not purchased until late in the 2009–2028 period do not incur recapitalization costs until some years later.

Cost Explanations for Programs

Army BCT (New)
Notional 20-Year Cost: $31.7B (FY 2009$). New Army BCTs are a program only in the Direct GWOT/COIN strategy. Adding six new Army BCTs constitutes the great majority of the Direct GWOT/COIN strategy's cost. This figure is not the cost of training, equipping, and supporting the roughly 3,500 soldiers in a BCT. Rather, it is one-sixth of

the total cost of adding 65,000 new soldiers to the Army.[1] This increase in Army active component force structure (along with the increase in Marine Corps force structure discussed below) was announced by the Bush administration in January 2007 (Garamone, 2007). It is understood to mean that the Army will stand up six new infantry brigade combat teams (Congressional Budget Office, 2007a, p. 12). The total cost of executing this plan was estimated by the Congressional Budget Office (CBO) in April 2007 (Congressional Budget Office, 2007a, p. 12). The CBO went beyond the cost of adding the roughly 3,500 troops in a BCT. Among other things, it assumed that the Army would preserve its current ratios of active duty soldiers to civilians and active duty ground troops to aviation units and added costs proportionately. Procurement costs also included military construction. Notably, the CBO's estimate of $70B in additional expenses for the Army from 2007–2013 (total, current-year dollars) was several billion dollars lower than the Army's own estimate (Congressional Budget Office, 2007a, p. 14).

The acquisition and O&S cost shares were also based on the CBO report. R&D costs were not included, as the new BCTs would not require unique types of equipment. The CBO provided the procurement (including military construction) and O&S costs for the additional soldiers between 2007 and 2013. The increase was expected to be fully executed in 2014. Procurement is treated as a one-time expense in the notional 20-year cost estimate. The O&S figure given for 2013 was assumed to be a steady-state annual O&S cost and was carried forward through 2028.

The timing applied to the new Army BCTs in the Direct GWOT/ COIN strategy is unique in that acquisition and O&S costs start immediately, in 2009. For every other program, the standing assumption is that some period of time must pass before money can be appropriated

---

[1] The administration's proposal would add 65,000 troops to the active duty end strength recommended in the 2006 QDR (Rumsfeld, 2006, pp. 400, 482). This end strength does not take into account the 30,000 Army soldiers temporarily authorized by the 2007 National Defense Authorization Act.

and spent, and then some further period must pass before the program in question can incur O&S costs.

Army BCT (Existing)

Notional 20-Year Cost: $28.3B (FY2009$). The cost of an existing Army BCT was derived from same CBO report described above. The key difference is that only O&S costs were considered; procurement costs were omitted, as these have already been incurred for existing units. As with above, this does not include any recapitalization costs.

It was assumed that BCTs could not be moved from one COCOM to another or cut from the force instantly. If the decision to cut a BCT was made in 2009, action would be taken in 2010 and savings would not appear until 2011. Similarly, large numbers of BCTs could not be moved or cut at the same time, but would have to be staggered over several years.

Train, Equip, Advise, Assist (TEAA) Initiative

Notional 20-Year Cost: $38.3B (FY2009$). The TEAA Initiative converts an Army BCT-equivalent into military training and advisory teams in the Build Local, Defend Global strategy. This program was inspired by two recent RAND reports (Grissom and Ochmanek, 2007; Gompert, Gordon, et al., 2008). Though those reports did not explicitly address the conversion of existing BCTs, they did put the capability gap in TEAA at roughly 5,000 people, which could staff between 400 and 500 MTTs.

As a BCT costs $28.3B in O&S, we added $10B over 20 years to account for the costs associated with conversion and the support structure needed to maintain more, smaller, geographically dispersed units.

The TEAA units were divided between CENTCOM and National Command. The TEAA initiative began to incur costs for these COCOMs in 2011, after time has passed to allow the appropriate training.

Note that although all the costs associated with the TEAA were charged to the Army, the other services could bear some of the burden or provide additional capability.

USMC Increase (27,000 Marines)

Notional 20-Year Cost: 76.5B (FY 2009$). The USMC increase, along with the six new Army BCTs described above, constitutes the great majority of the cost of the Direct GWOT/COIN strategy. As with those new BCTs, the USMC increase is based on the Bush administration plan to increase the size of the ground forces, and the cost data were likewise drawn from the CBO report on the subject.

Unlike the Army increase, the principal aim of which is to add six infantry brigade combat teams, there was no clear indication of how the additional Marines were to be integrated into the Marine Corps structure. The increase was therefore treated as a single cohort of 27,000 additional Marines over the active duty end strength recommended in the 2006 QDR.

The CBO estimated the cost of 27,000 additional Marines based on the cost to equip and operate a spectrum of units, from infantry battalions to fighter aircraft squadrons to support companies (Congressional Budget Office, 2007a, p. 12). No research and development costs were included, as existing equipment types would be used to outfit these units. Procurement costs, which also include military construction, came to roughly 20 percent of the total 20-year cost. Procurement was treated as a one-time cost. O&S costs made up the balance.

It was estimated that it would take three years to acquire the equipment for these 27,000 Marines; the CBO shows additional procurement costs dropping to near zero by 2012. Because the first of these additional Marines will be serving long before the last joins the force (and indeed some portion of these Marines are already in the force today), O&S costs were begun in 2010, before acquisition is complete. Although this is imprecise, the larger-than-appropriate 2010 and 2011 O&S bills compensate for an O&S bill of zero in 2009.

Green Water Squadron

Notional 20-Year Cost for First Unit: $6.5B (FY 2009$), Notional 20-Year Cost for Subsequent Units: $4.4B (FY 2009$). The green water squadron is a concept that draws on research done at the National Defense University (NDU) for the former Office of Force Transformation in the Office of the Secretary of Defense. That research developed alternative future Navy fleet architectures and explored the benefits of using larger numbers of small vessels than exist in the fleet today (Johnson and Cebrowski, 2005). This study formed the basis for war games sponsored by the Vice Chief of Naval Operations. Among the results of the games was the utility of a green water squadron, which drew both on existing ships and on ship types from the NDU study to form a unit optimized for operations in restricted waters (littoral, riverine, and straits). A green water squadron comprises one LPD-17 amphibious warfare ship, one Littoral Combat Ship (LCS), and two types of notional future ships from the NDU study:

- a 13,500-ton aircraft carrier (one per green water squadron) for vertical/short take-off and landing aircraft and unmanned aerial vehicles (UAVs) called an X-CRS
- a 1,000-ton surface combatant (six per green water squadron) called an SSC-1000.

The cost to develop, procure, and operate these nine ships was estimated largely on the basis of data drawn from an Institute for Defense Analyses (IDA) report on which the NDU study also drew. IDA estimated the average unit procurement costs to be the following (in FY 2005$):

- $780M for LPD-17
- $400M for LCS
- $250M for X-CRS
- $150M for SSC-1000 (Greer, 2005).

Annual O&S costs (also in FY 2005$) were estimated to be the following:

- $36M for LPD-17
- $6.5M for LCS
- $20M for X-CRS
- $3.3M for SSC-1000.

Converting these figures to FY 2009 dollars yielded a combined procurement cost and 20-year O&S cost for each squadron of $4.4B.

However, this $4.4B does not include the R&D costs necessary to develop two new types of ships. The estimated cost to develop the X-CRS and the SSC-1000 is based on the cost to develop one of the Navy's latest ships, the LCS. The LCS program received nonconstruction-related research, development, test, and evaluation (RDT&E) funding on the order of $750M.[2] This figure is rounded up to $1 billion and added as an R&D cost for both the SSC-1000 and the X-CRS, meaning that the first green water squadron would cost an additional $2B (for a total, after rounding, of $6.5B).

It was estimated that the Navy would need some years to develop two new ship types and build the many vessels called for by the green water squadrons (eight squadrons in the Build Local, Defend Global strategy and seven in the Respond to Rising China strategy—or 72 and 63 ships, respectively). Timing for the squadrons was estimated on the basis of what was deemed theoretically feasible, but with an acknowledged bias toward getting these units into the force (that is, showing their resource implications) as soon as possible. R&D for the first squadron was estimated at five years, and procurement for all squadrons was estimated at three years. Procurement of squadrons was staggered, as the Navy certainly could not buy all these ships at once.

SSGN Conversion (Two Boats)

Notional 20-Year Cost: $2B. The SSGN program converts nuclear missile submarines (SSBNs) into boats able to carry conventional cruise missiles and special operations forces. Current plans call for conversion of four SSBNs that were otherwise due to be retired. The cost per boat is roughly $1B. The conversion process takes about three years. The posited conversions are in addition to those already programmed and do not draw on SSBNs that would otherwise be retired; the SSBNs in question would otherwise continue to serve in the force. It was assumed

---

[2] See a recent Congressional Research Service study (O'Rourke, 2007). In an unusual action, the LCS program funded construction of initial ships and modules from the RDT&E account.

that the cost to operate an SSGN is about the same as for an SSBN, so no additional O&S costs were added to the strategy.

DDG-1000/CG(X) (Various)

Notional 20-Year Cost per Ship: $4.7B. The resource implications for the DDG-1000 and CG(X) should be roughly the same; the Navy plans to base the latter substantially on the design of the former but add area air defense capability while taking away strike capability. The differences between the two types of ships may blur, however, and either could be supplanted by a third (or fourth) type of ship in the same general class. For illustrative purposes, therefore, we were not specific about the types of future capital ships included in the Respond to Rising China strategy, and we treated DDG-1000/CG(X) as a single program. Although the DDG-1000 program is further along, CG(X) cost data were the main basis for the cost estimate, in part because the strategy calls for these ships to provide air and missile defense.

CBO estimated the average unit procurement cost of the CG(X) at $3.9B (FY 2007$) (Congressional Budget Office, 2007c, p.16). The DDG-1000/CG(X) was assumed to already be part of the Navy's baseline, so the additional ships do not include R&D costs (Gilmore, 2005).[3] CBO estimated the O&S costs for DDG-1000 at between $25M and $32M annually (FY 2007$) (Gilmore, 2005, p. 5). A $32M annual O&S cost was used to arrive at a total program cost (procurement plus O&S) of $4.7B per ship (FY 2009$). For ease of analysis, these ships were packaged into programmatic packages of two ($9.4B) or six ($28.2B).

The timing of the DDG-1000/CG(X) units was tied to an estimate for the schedule of the CG(X). The first CG(X) is due to be procured in FY 2013. It was estimated that it would take two years to build a two-ship unit and five years to build a six-ship unit. In the case

---

[3] The CBO's average unit procurement cost encompasses all costs associated with the CG(X) program, including R&D. We elected to use this figure despite the fact that the ships represented in the strategy are not considered to bear R&D costs. This (small) cost increase was deemed acceptable because the CBO indicated that its estimate of $3.9B might be optimistic; historical examples suggest that the Navy may not realize the expected savings from using a common hull for the DDG-1000 and CG(X).

of six-ship units, O&S is begun before the last ship in the notional unit is complete, in order to capture (roughly) the costs associated with those ships produced first and already in the fleet.

As discussed at the beginning of this appendix and alluded to in the section on the green water squadron, the strategies were developed without detailed consideration of nonmonetary resource implications, shipyard capacity being the most prominent.

Medium-Range Bomber Wing

Notional 20-Year Cost: $68.6B (FY 2009$). The cost estimate for a future medium-range bomber was based on a 2004 RAND analysis of alternatives for a next-generation gunship (Moore et al., 2004). Although a medium-range bomber and a gunship are different, the favored gunship alternative in the RAND study could, with modifications, meet both missions. In fact, the gunship study team indicated that one of the preferred alternative's attributes was that it could be adapted from a future bomber airframe.

The notional plane in question is a subsonic aircraft with a large payload and a combat radius of about 2,000 miles. The gunship study indicated a program R&D cost of about $12B (FY 2007$), a unit procurement cost of $370M (FY 2007$), and an annual O&S cost per plane of about $10M (FY 2007$).

These numbers were used to estimate the development, procurement, and operation of a 72-plane bomber wing. Since this medium-range bomber program is not in the Analytic Baseline, the strategy had to bear the burden of R&D, which was estimated to be the same as for the gunship: $12B. Unit procurement cost was also estimated to be the same: $370M per plane. One-hundred aircraft were procured to keep 72 flying—the additional 28 aircraft are for training, testing, and reserve.[4] O&S costs for this program total about $15B over 20 years.[5]

---

[4] The rough planning factors applied for keeping 72 aircraft combat-ready are as follows: two test aircraft for the new program, plus an additional 10 percent of the combat-ready inventory for training, plus about 10 percent of the flown inventory (combat plus test plus training) for backup, plus a further 10 percent to replace expected peacetime losses.

[5] O&S costs are applied only to the flown inventory—combat aircraft plus test aircraft plus training aircraft, or 81 planes in this case.

This works out to a 20-year total (in FY 2009$) of $59B. The share of that total for R&D, acquisition, and O&S is roughly 20 percent, 60 percent, and 20 percent, respectively.

The medium-range bomber wing is acquired over four years starting in 2015 and starts accruing O&S costs in 2020. This estimate is based on the Air Force's targeted IOC for a long-range surveillance and strike aircraft: 2018. Without regard to whether the Air Force would or could pursue two bomber programs at once, the 2018 time frame is a reasonable notional goal for fielding a new type of bomber. Note that some portion of the medium-range bomber wing would be flying before acquisition of all 100 planes was complete, so some O&S costs have gone uncounted.

Long-Range Surveillance and Strike Aircraft Squadron

Notional 20-Year Cost: $25.4B (FY 2009$). It was assumed in this case that the Air Force will procure some number of long-range surveillance and strike aircraft, per the Air Force leadership's announced plans (Sirak, 2007). Because the strategies are drawing on a program projected to be in the Analytic Baseline, the squadron bears no additional R&D costs.

This plane will be stealthy, subsonic, and manned. RAND staff with knowledge of the Air Force's program estimate the per-plane acquisition cost at $500M.[6] The squadron consists of a total buy of 33 additional aircraft. With a target of keeping 24 aircraft combat-coded, roughly 10 percent more training aircraft, 10 percent more reserve aircraft, and 10 percent more attrition aircraft are added.[7] That leads to an acquisition cost of $16.5B.

Given an O&S estimate of $50,000 per flight hour and the B-2's annual total of about 330 flight hours, a 20-year O&S cost of about

---

[6] This tracks roughly with a 2005 CBO estimate, which placed the per-plane cost of a long-range bomber at about $430M (FY 2007$, not including R&D costs) (Congressional Budget Office, 2006).

[7] As with the medium-range bomber, this likely underestimates the additional planes needed to keep 24 combat-coded. Unlike the medium-range bomber program, however, in this case an (assumed) existing fleet of long-range bombers can be called on for training. No additional test aircraft are needed.

$9B was projected. Expressed in FY 2009$, the total notional 20-year cost for the long-range bomber squadron is $25.4B.

The Air Force intends to achieve IOC for the long-range surveillance and strike aircraft in 2018. On that basis, acquisition was begun in 2015 and O&S costs started to accrue in 2019.

Long-Range Missiles

Notional 20-Year Cost: $1B (FY 2009$). The cost estimate for long-range missiles is based on Navy plans to fit existing Trident submarine-launched ballistic missiles with nonnuclear penetrating warheads. This program would equip each of the Navy's patrolling Trident submarines with two conventionally equipped missiles in addition to 22 missiles with nuclear warheads. The Navy requested just over $500 million in FY 2007, to be spread over five years, for the so-called conventional Trident modification (Woolf, 2007). This level of funding would allow for deployment of the missiles in 2012. The cost estimate used in this report arbitrarily increased this funding level to $1 billion to allow for all other aspects of 20-year costs.

UAV Squadron or Detachment (HALE)

Notional 20-Year Cost: $5.6B (FY 2009$) for a Squadron, $1.2B for a Detachment (FY 2009$). The HALE UAV was assumed to be the existing Global Hawk program. The strategies used two different HALE UAV–based units. A squadron was assumed to have a total of 24 aircraft. To calculate O&S, a detachment was assumed to have five aircraft, with (for O&S purposes) only four ready to fly. The strategies incurred no R&D costs, as the Global Hawk has already been developed.

The acquisition cost per aircraft, inclusive of ground stations and support, is about $100M (FY 2007$).[8] No additional payload costs were included, although some could result if the aircraft were to be used as a limited replacement for satellite ISR, communications, or GPS capability, as the strategies suggest. The annual O&S cost per

---

[8]   U.S. Air Force (2007), p. 4/45.

aircraft was estimated to be about \$7M.[9] Of the 24 aircraft in a squadron, it was assumed that 16 of these would be mission-ready, with four aircraft for training and four for backup and attrition, and so O&S was calculated for 20 aircraft. For a detachment, O&S was calculated for four aircraft.

It was assumed that the Global Hawk production line could accommodate the additional aircraft (15 in the Build Local, Defend Global strategy, 34 in the Respond to Rising China strategy) in the relatively near future. O&S for all new HALE UAV units starts in 2012.[10]

UAV Squadron (MALE)

Notional 20-Year Cost: \$0.7B (FY 2009\$). The estimated cost for a medium-altitude/long endurance (MALE) UAV squadron was based on the Predator-B (also known as the MQ-9 Reaper), the new, larger, heavily armed version of the MQ/RQ-1 Predator UAV. UAVs in the Build Local, Defend Global strategy would not necessarily have to be configured as the Predator-B, but that aircraft's size and capabilities were a reasonable stand-in for some mix of advanced ISR and strike capabilities.

The Predator-B is estimated to cost about \$10M per aircraft, exclusive of R&D but inclusive of ground stations and support. Thirty-four aircraft were procured for each squadron: 24 to be combat-ready, plus an additional 10 percent for training, an additional 10 percent for backup, and an additional 20 percent for attrition. The total procurement cost per squadron was just under \$350M. For O&S, it was estimated that the MALE UAV squadron would incur roughly the same ratio of procurement cost to 20-year O&S cost that Global Hawk did: about 50:50. That doubled the notional 20-year cost of a MALE UAV squadron to just under 0.7B (FY 2009\$).

It was assumed that the Air Force and Navy could procure, and that industry could produce, one additional 34-plane MALE UAV

---

[9] Based on the Selected Acquisition Report.

[10] Global Hawks are being procured at a rate of about five per year, with a total planned buy of a little over 50, so these are substantial additions to the force.

squadron a year. The Build Local, Defend Global strategy has O&S costs for the first new MALE UAV squadron starting in 2011.

C-17 Squadron

Notional 20-Year Cost: $4B (FY 2009$). The C-17 squadron is assumed to include 12 aircraft. The Build Local, Defend Global strategy shifts existing C-17s from one COCOM to another, so only O&S costs are considered. The Selected Acquisition Report for the C-17 indicates that the annual per plane O&S cost is $16M (FY 2007$). The notional 20-year cost for a 12-plane squadron is $4B (FY 2009$). It was assumed that those planes will be moved by 2011.

Special Operations Forces (Soldiers)

Notional 20-Year Cost per Group (Existing): $4.1B (FY 2009$); Notional 20-Year Cost per Group (New): $4.4B (FY 2009$); Notional 20-Year Cost per Battalion (New): $2.2B (FY 2009$); Notional 20-Year Cost per Company (New): $0.3B (FY 2009$). Several unit types were used to express SOF investments: groups, battalions, and companies. These unit types were convenient signifiers of the scale of resources that we intended to commit to SOF in a given COCOM. Further, for the sake of simplicity, all SOF were assumed to be active duty Army units. As such, the various groups, battalions, and companies should be considered representative of cost implications but not reflective of current force structure; the Build Local, Defend Global strategy, for instance, nominally moves more active duty Army SOF groups than currently exist. Further, actual SOF forces would include units from the other services.

The cost estimate for the various SOF units was based on data from the Army's FORCES Cost Model (FCM). The FCM placed the acquisition cost for a new group (~1,350 men) at about $260M. The annual O&S cost for a group was $200M. The FCM placed the acquisition cost for a battalion (~440 men) at $80M and the annual O&S cost at $65M. The acquisition cost for a SOF company was $10M and the annual O&S was $15M. These figures were used to calculate the notional 20-year cost of the units. Existing units did not reflect acquisition costs.

It was assumed that SOF units can be shifted from COCOM to COCOM relatively quickly, so O&S costs for all units start in 2011.

SOF Trainers (Company)

Notional 20-year Cost: $0.2B (FY 2009$). The Build Local, Defend Global strategy made a distinction between SOF troops needed for full-spectrum operations, from combat through civil affairs, and SOF troops whose mission would almost exclusively be to train local security forces. It was assumed that such training forces would be equipped and trained differently and less expensively. The 20-year cost of a training company was estimated at 75 percent of the cost of a regular SOF company.

Enhanced National Missile Defense

Notional 20-Year Cost: $134B (FY 2009$). A 2004 CBO report on alternatives for boost-phase interception of ballistic missiles provides the basis for a rough estimate of the cost of a more capable national missile defense system (Congressional Budget Office, 2004). The CBO addressed several alternatives to counter a limited number of launches from Iran and North Korea. We had in mind a capability to cope with a limited number of launches from China, a more sophisticated adversary with a large territory that complicates boost-phase intercepts. We therefore selected the most ambitious (and most costly) program detailed by the CBO: a space-based constellation of interceptors with some ability to intercept faster-burning solid-fueled ballistic missiles. The 20-year cost, in FY 2009$, was $134B. The CBO indicated that about 10 percent of that would be for R&D, 30 percent for acquisition, and 60 percent for O&S.

It was estimated that R&D would start in 2009 and the system would be fielded beginning in 2022.

Security Assistance and Foreign Assistance

Notional 20-Year Cost(s): Varied. The security assistance and foreign assistance cost estimates were derived in part from a recent RAND report on counterinsurgency (COIN) and in part from current U.S. funding patterns. The RAND report suggests improvements in

U.S. COIN capability in several areas (Gompert, Gordon, Grissom, Frelinger, Jones, Libicki, O'Connell, Stearns, and Hunter, 2008). One is to greatly enhance the nation's civil capabilities for COIN. Adding 5,000 to 10,000 civilian personnel (mostly to the U.S. Agency for International Development and the Department of State) would help greatly in this regard, at an annual cost of between $2B and $4B. We used this figure as a starting point for estimating the cost of security assistance in the Direct GWOT/COIN and Build Local, Defend Global strategies. The Direct GWOT/COIN strategy, with its focus on direct U.S. military action, placed less emphasis on civil capabilities and so executed about $2.5B in annual security assistance. The Build Local, Defend Global strategy, with its heavy emphasis on nonmilitary activity, spent over $3B a year on security assistance. In addition to spending for more U.S. civilian personnel, monies were made available to contract further expertise and fund limited capital investment in foreign security–related infrastructure.

In the Build Local, Defend Global strategy and the Respond to Rising China strategy, there are also foreign assistance outlays. The RAND report estimated that between $10B and $15B a year in additional U.S. assistance (plus additional assistance from allies) would be required to prevent insurgencies from developing and for building economic, technical, and political capacity in target nations (Gompert, Gordon, Grissom, Frelinger, Jones, Libicki, O'Connell, Stearns, and Hunter, 2008). This estimate was used to guide total annual spending in the Build Local, Defend Global strategy; the total was divided among the COCOMs roughly in line with current U.S. foreign operations spending (exclusive of much funding in support of operations in Iraq and Afghanistan), with about 60 percent going to CENTCOM and about 20 percent each going to PACOM and AFRICOM. In the Respond to Rising China strategy, the spending is less ambitious but is undertaken with the same aims in mind: The money is still intended to build the economic and governance capacity of partner nations.

Treatment of Risks

Portfolio analysis should identify and characterize both the upside and the downside potential of the strategies it compares (Davis, Kulick, and Egner, 2005). In this monograph, we deal only with the downside, i.e., with risks.

Describing risks is nontrivial because of their diverse character and the ambiguity of the meaning of "risk" in the English language, including its conflation with "uncertainty." This appendix describes the issues from our perspective and how we chose to address risks in Chapter Five.[1]

Prior Definitions

Let us first review some of the common definitions of risk, which are not adopted in this monograph.

---

[1] This appendix benefited from discussions with and suggestions from Lynne Wainfan, who is currently writing a Ph.D. dissertation in the Pardee RAND Graduate School. A huge and sometimes disputatious literature exists on risk management, much of which can be found cited in a recent National Research Council study responding negatively to proposed guidance on the subject by the Office of Management and Budget (National Research Council, 2007). Revised guidance was issued recently by OMB and the Office of Science and Technology Policy (Dudley and Hays, 2007). Our concerns are more narrow and analytic, but many of the subtle issues of policy and process arising in the debate about OMB guidelines have analogues in defense planning.

The Mathematician's Preferred Definition

A rigor-enhancing definition that was introduced almost a century ago regards risk as a class of uncertainty for which the underlying probability distribution of outcomes is known, as when making bets on the flip of a coin. That definition was adopted in early texts on systems analysis (Madansky, 1968) and can frequently be seen today in technical literature, but it fits poorly with normal English.

Risk as a Product or Expected Value

Another common definition regards risk as the product of a likelihood of a bad outcome and the negative consequences of that outcome. Although consistent with the natural-language sense that "risk" depends on both likelihood and consequences, the formulation as a product is too narrow and can be counterproductive. A number of problems are common. First, many risks are associated not with single events and likelihoods but with a number of possible events of varying likelihood. This implies the need for something more complex, such as a weighted sum or an integral formulation, e.g.,

$$R = \bar{E} = \int P(x)C(x)ds,$$

where P(x) is the probability density of an outcome x with a consequence C(x), and risk R is regarded as the so-called "expected value" (actually, the mean). This definition is common among economists and mathematicians.

Even with the expected-value generalization, there are problems. Some risks are associated with potential events having "infinite cost," such as the consequences of general nuclear war. In such cases, people may still see the risk as low (or at least act as if it is). This may be because of a cognitive shortcut in which we treat probabilities below some threshold as effectively 0, or because we implicitly discretize consequences. To illustrate the latter, suppose that we measure consequences and risk as values from the set {1,3,5,7,9} (equivalent to using a five-color scorecard) with 9 being the worst. If the perceived likelihood is less than 0.17, the expected value is 1, the lowest possible—even though 0.17 (about 1 in 6) is not especially small.

Paradoxically, some risks that involve low probabilities and high but not infinite consequences are considered serious enough to merit housing codes, insurance, major defense expenditures, or other protective measures, even though the expected-value formulation of risk is small (Haimes, 1998). It would seem that people implicitly fold into assessments of risk an implicit notion about whether anything can be done about the threat. If not, the risk is ignored, often without conscious recognition of doing so. If something can be done, however, the risk is recognized. A special case of this is the empirical fact that people will typically reject bets that have the potential to cost "too much," even if an economist—looking at expected return—would regard the bet as highly favorable.[2] By and large, we do not "bet the farm" based on expected value. This may seen as a cognitive bias or a wise recognition that we live only once.

For reasons that should become clear in the following section, neither of these two common definitions of risk—as an uncertainty with a known probability distribution, or as the expected value of a likelihood-consequences product—are suitable for our purposes.

Definition and Classes of Risk in This Monograph
The perspective we take in this monograph is that:

> *"Risk" is a measure of those negative consequences of uncertainty that can be recognized and are appropriate to account for.*

In this definition, "risk" is shorthand for what might be called the "relevant downside of uncertainty." The last phrase reflects the attempt to be realistic. It is desirable to include as much risk as one can imagine in calculations and to be creative in doing so, but some risks will be missed and some can be ignored by choice.

---

[2]   To be sure, utility could be generalized here to include a term for risk, in which case the bets in question would have a much lower expected utility and thus be unattractive.

Reference to "negative" consequences is appropriate because uncertainties can work in both directions and, as discussed in other work, accounting for upside potential is sometimes essential to offset the bias introduced by dwelling only on risks. This is true in both strategic planning and operations planning (Davis, Kulick, and Egner, 2005; Davis and Kahan, 2007). Our definition of risk also means that something carries with it no risk if it has zero likelihood or if, even if it occurs, it makes no difference. It acknowledges that we are not talking about some things that might have been included, but are not—the risk of a not-yet-detected comet or meteor striking the earth, of general nuclear war in the absence of a major-power crisis, etc.[3] Beyond that, the definition is deliberately less precise than specifying a product or integral formulation. This is important in our context because DoD planners are obligated to manage risks generally, not just the nominal values of various risks. Low-probability, high-consequence events must be addressed, unless they involve matters over which the United States has absolutely no influence. Even then, they should be considered to better understand the significance of addressing risks that can be addressed.

Before identifying the classes of risk that we consider, we first note an implication of the above definition for PAT-related analysis. Suppose that we are comparing options against three measures of goodness (e.g., expected consequences for the health of three combatant commands). Suppose further that the summary assessments are good, good, and bad for the three categories. We do not consider the bad as indicating "risk." Rather, bad is the nominal evaluation itself with nothing said about uncertainty. Just as we do not say that there is risk if one jumps off a high building (there is the expectation of death, not

---

[3] In the words of former Secretary of Defense Donald Rumsfeld, "As we know, there are known knowns. There are things we know. We also know there are known unknowns. That is to say, we know there are some things that we do not know. But there are also unknown unknowns. The one we don't know we don't know. And if one looks throughout the history of our country and other free countries, it is [those in] the latter category that tend to be the difficult ones" (Defense Link, February 12, 2002 transcript). Although often mentioned humorously, Rumsfeld's comments were quite true.

a risk), so also if the evaluation is bad, then we need say nothing about risk per se.

A confusing subtlety is that if the evaluation is the result of looking at several lower-level factors, which may include risks, then—although the assessment itself does not measure risk—its value might be driven by risks (e.g., if the evaluation were to take the worst of the component results and the component called *risks* was the worst). Some of the possibilities can be seen in Figure E.1, which postulates a portfolio analysis producing a summary assessment based on three measures, one of which is overall risk, with each measure being composed of subordinate components and with overall risk being determined by the risk components of the other measures.[4]

Figure E.1
Composite, or Overall, Risk

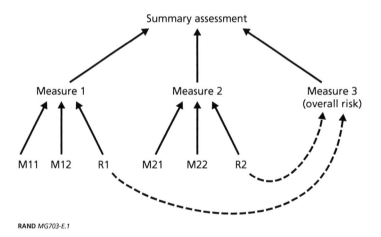

RAND *MG703-E.1*

---

[4]   From a mathematical viewpoint, Figure E.1 is worrisome: R1 and R2 are "double-subscribed" in that they contribute to the summary assessment in two different chains. Is that redundant? Conceptually, it is not, but in some analytical contexts there might be some problems when calculating a composite or summary assessment, in which case the analyst might wish to give no weight to the equivalent of Measure 3, treating it as a separate issue to be displayed but not as one to be rolled into effectiveness calculations.

With this background, we have chosen in this monograph to use the term "risks" in five relatively distinct ways, as summarized in Table E.1.

Table E.1
Classes of Risk

Classes of Risk	Description	Representation in PAT
Case-related	Risk depends on test cases, e.g., scenarios of future or war scenarios, or combinations of parameter values.	This may be explicit (i.e., a top-level variable called risk) with drill-down. Alternatively, a summary risk factor might be based on the fraction of higher-resolution test cases for which results are adverse.
Assessment	Reality may be more adverse than assessment because of errors in estimating, e.g., capabilities, behaviors, consequences, or future salience of an issue to future leaders.	Alternative extended perspectives are possible, with different value weightings, priorities, requirements, and evaluation algorithms.
Composite or overall	A summary measure of an option's goodness that characterizes overall risk, recognizing risks in each of the other summary-level measures.	This is an explicit summary-level measure based on component risks of other measures.
Aggregation	A summary assessment may obscure, understate, or overstate risks identified at higher levels of detail, perhaps by misestimating probabilities, glossing over "bimodal" distributions, or evaluating the aggregate as the worst of the components.	This features drill-down capability and warning flags.
Inherent	This is a type of risk for which we do not have (and might not benefit from) more fine-grained assessment. This might be riskiness from uncontrolled and perhaps unobservable factors, e.g., risk of development failure when cutting-edge technology is involved, of environment-shaping efforts from random events causing negative misperceptions, or failure of local would-be partners' political systems.	Contains explicit recognition (e.g., development risk for a weapon-system option) and warning markers, perhaps applied to an entire category (e.g., any evaluation of outcome in dealing with North Korea might be uncertain, with a large downside risk).

Case-Related Risk (Variation)

Suppose that we are evaluating an option with a particular measure (e.g., the health of PACOM as reported by the COCOM after considering a variety of situations, including capacity for shaping operations and warfighting). Suppose that the COCOM's evaluation was based on analysis of a number of different cases. The cases might reflect different assumptions about the future security environment in Asia, details of warfighting scenarios, or something else. That is, suppose that we have constructed a "scenario space" in the broadest sense of the term "scenario," and the health of PACOM varies depending on where we look in the space—assuming considerable variation of results across cases. Case-related risk might then be measured by the relative plausibility or importance of the portions of the scenario space/case space for which results would be adverse—below some threshold of acceptability.[5]

Assessment Risk

As a separate matter, consider the evaluation itself for a well-defined case. That assessment might be overly optimistic: A realistic evaluation might be much more adverse by giving adversaries more credit for future capabilities; or it might make more alarmist assumptions about how an adversary would react to a U.S. military action (perhaps with an irrational use of nuclear weapons in what the adversary would see as a last-gasp strike at his enemy). Or it might make more pessimistic assumptions about the adversary's capabilities. For the context of this monograph, other examples are particularly salient. An evaluation might anticipate success for a Direct GWOT/COIN strategy if merely some level of resources were provided; however, that assessment could underestimate the nationalistic reactions to U.S. involvement in the Middle East and the campaign could prove counterproductive. The Build Local, Defend Global strategy might be seen as likely to be effective if merely the United States provided the foreign assistance that

---

[5] DoD analysts sometimes predict outcomes in some test scenarios and then refer to the difference in results between the nominal case and the worst case as a measure of risk. This is misleading if the model itself is unreliable and potentially optimistic or the test cases exclude plausible cases where results would be more adverse.

the strategy assumes, as well as the military component. The reality might be that local would-be partners would be so corrupt or inept as to doom any such strategy. The Respond to Rising China strategy is intended to "avoid a vacuum" but could trigger an escalated arms race and a more aggressively emergent Chinese defense posture.

In principle, these types of assessment risk could all be covered in the case-related risk category, but in practice it may be useful to separate them. Assessment risk is intended for when it is useful to highlight seriously different attitudes, beliefs, or value systems. This may be so because the issues involved are strategic and appropriate for discussion with senior leaders. A historical example of such an issue was, at the time of the Vietnam war, whether the war should be seen as invasion by an aggressor who could be persuaded to desist by cost-benefit calculations or as a civil war by a fiercely nationalistic movement that would tolerate enormous pain in pursuit of its objectives and continue to do so for years if necessary.

Within PAT, such uncertainties are treated by the alternative perspectives and "extended perspectives" described in Chapter Five. These may be defined analytically with different weightings across categories, different combining rules (aggregation rules), and seriously different qualitative assumptions about the consequences of actions. Such perspectives may represent not just current decisionmaker values and judgments but also, for example, potential future national attitudes that should be anticipated.[6]

Composite Risk

Composite risk is a composite measure determined by component risks (see Figure E.1). In this monograph, we have treated it as a top-level measure.

---

[6] As an example, consider planning for ballistic-missile defense. At a given time, political-level priority might be given to defense against an accidental launch, a small attack by a rogue state, or a full-up defensive shield. Whatever the priorities, planners should anticipate that they will change as the result of international events or nonevents. Thus, they may wish to do alternative assessments of proposed programs with that in mind.

Aggregation Risk

Aggregation risk is the risk that the process of aggregation—so essential to the preparation of summary assessments—may obscure problems or even misrepresent the situation. Aggregation necessarily involves judgments, such as whether to average results across submeasures or, for example, to characterize the aggregation by the worst result at the lower level. In PAT, where drill-down has been provided, this issue is dealt with explicitly; essentially, the issue is handled by recognizing different cases. However, where it is not, warning flags can be used to alert the viewer to important assumptions, bimodal lower-level results, or other complications.

Aggregation risk is illustrated in Figure E.2, which assumes that analysis has evaluated results for six cases, but that it is now necessary to simplify and summarize. As a first step, the analyst may discard bad cases regarded as below some level of likelihood. The result (center) is a smaller set of test cases, results for which are bad for only one. If a further simplification is necessary, the aggregation rule might characterize the result as marginal (yellow). There is a risk that either of the aggregations will prove erroneous and overestimate what is being assessed. The

Figure E.2
Case-Related Risk

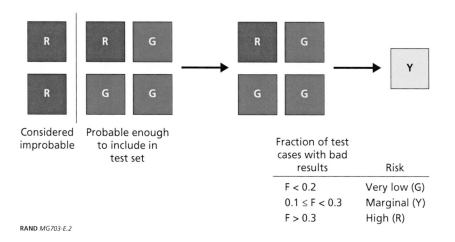

| Considered improbable | Probable enough to include in test set | | | |

Fraction of test cases with bad results	Risk
F < 0.2	Very low (G)
0.1 ≤ F < 0.3	Marginal (Y)
F > 0.3	High (R)

discarded cases may turn out to be important, but even if they do not, conveying the sense of "marginal" (yellow) for the most aggregated result is problematic because in this case it is reflecting a near-even mix of good and bad results (in other cases, it might mean that we expect results to be so-so, in which case yellow would be appropriate). Results are never expected to be marginal. Were we to be treating analogous issues in statistics, the admonition would be to show a measure of variance or a measure of the "probability" of adverse results, in addition to showing the summary.

Inherent Risk

Finally, inherent risk is a category for which we do not have (and might not benefit from) more fine-grained analysis. Attempting to build a new weapon system dependent on certain types of cutting-edge technology involves high risk, which can be asserted without decomposing the problem further. Attempting to "improve" results with more detail may be counterproductive because the clearly identifiable risks that can be evaluated may seem more manageable than those that are less clear-cut. NATO's strategy of flexible response during the Cold War had obvious inherent risks.

As a second example, consider that decades after the event, we now know that the Soviet military in Cuba not only had nuclear weapons but had been given predelegated authority to use them in the event of an invasion. The delegation of authority might well be considered to have been one of the significant "unknown unknowns" of the crisis, to use the language of Secretary Rumsfeld mentioned above. Fortunately, top U.S. and Soviet leaders were well aware of inherent risks. As a result, they were more cautious than some of their subordinates (most of Kennedy's executive committee of advisors [EXCOM] favored invading Cuba).[7]

---

[7]   A primary source is the set of audiotapes of Kennedy's EXCOM discussion during the Cuban Missile Crisis; they are available from the National Security Archive of the George Washington University, which can be readily found with a Google search.

Additional Concepts and Subtleties

It is a core element of portfolio-management approaches that one may be willing to tolerate mediocre results or higher risks in one portion of the portfolio to achieve better results elsewhere. If one chooses an option that has such features, then one is accepting an implied risk. There is nothing particularly subtle in this, but recognizing the implied risks—and taking them seriously—is obviously important.

Some of the risks that we have described are not so obvious and become visible to the decisionmaker only with some effort. The value of making an effort to do so is well described in a book on assumption-based planning (Dewar, 2003). It describes methods for identifying the underlying assumptions of a plan that are both important and potentially wrong. It refers to those as the plan's vulnerable assumptions. The same concept can be used when discussing assessments of strategy.

A properly constructed analysis embedded in PAT can help do so. For example, the warning markers can highlight important assumptions or the existence of highly adverse special cases that have been glossed over. Also, the basis for the various aggregations can be displayed, giving the decisionmaker more visibility into what lies beneath the summary analysis.

The Units of Risk

Because of the many different meanings of "risk" and the difficulty of defining it precisely and well, it can be measured in different ways depending on context. In effect, a given risk variable may be evaluated essentially as

- the likelihood of unacceptable consequences
- the consequences of the worst plausible case among cases plausible enough to be taken seriously
- the expected value of outcomes worse than the nominal outcome, in comparison with the nominal outcome.

This variation of effective meaning may be disconcerting, but each of the evaluations is a context-dependent approximation of something complicated. Pragmatically, for the purposes of strategic planning (as distinct, say, from comparing portfolio risks in personal finance or from comparing development options with components having different technical maturity levels), we cannot avoid the heuristic reasoning unless we oversimplify the problem by ignoring aspects of risk that do not lend themselves easily to mathematics. The reader should assess reasonableness by looking at concrete examples in context (e.g., as in Chapter Five).

Bibliography

Alberts, David, and Richard Hayes, Understanding Command and Control, Washington, D.C.: Command and Control Research Program, Department of Defense, 2006.

Bexfield, James, "A State of the Analysis Community," a briefing, May 9, 2007. As of July 24, 3008: www.msco.mil/files/DMSC/2007/Bexfield_AnalysisCommunityOverview.ppt

Bracken, Paul, "Managing to Fail: Why Strategy Is Disjointed," *The American Interest,* Vol. 3, No. 1, September–October 2007.

Bush, George W., *National Security Strategy of the United States,* Washington, D.C.: White House, 2006.

Chairman of the Joint Chiefs of Staff, *Instruction on Capabilities-Based Planning,* Washington, D.C.: Joint Chiefs of Staff, CJCSI 3000.01, 2007.

Cheney, Richard, *Defense Strategy for the 1980s: The Regional Defense Strategy,* Washington, D.C.: Department of Defense, 1993.

Chu, David C., and Nurith Berstein, "Decisionmaking for Defense," in Stuart E. Johnson, Martin C. Libicki, and Gregory F. Treverton, eds., *New Challenges, New Tools for Defense Decisionmaking,* Santa Monica, Calif.: RAND Corporation, 2003. As of April 17, 2008: http://www.rand.org/pubs/monograph_reports/MR1576/

Cohen, William, *Report of the Quadrennial Defense Review,* Washington, D.C.: Department of Defense, 1997.

Congressional Budget Office, *Estimated Costs of a Potential Conflict with Iraq,* Washington, D.C.: Congressional Budget Office, Congress of the United States, 2002.

———, *Alternatives for Boost-Phase Missile Defense,* Washington, D.C.: Congress of the United States, 2004.

———, *Alternatives for Long-Range Attack Systems,* Washington, D.C.: Congress of the United States, 2006.

———, *Estimated Cost of the Administration's Proposal to Increase the Army's and the Marine Corps's Personnel Levels,* Washington, D.C.: Congress of the United States, 2007a.

———, *Evaluating Military Compensation,* Washington, D.C.: Congress of the United States, 2007b.

———, *The Navy's 2008 Shipbuilding Plan and Key Ship Programs,* Statement of J. Michael Gilmore and Eric J. Labs Before the Subcommittee on Seapower and Expeditionary Forces, Armed Services Committee, U.S. House of Representatives, July 24, 2007, Congress of the United States, 2007c.

Datar, Vinay, and Scott Matthews, "European Real Options: An Intuitive Algorithm for the Black-Scholes Formula," *Journal of Applied Finance,* Vol. 14, No. 1, 2004.

Davis, Paul K., "Planning Under Uncertainty Then and Now: Paradigms Lost and Paradigms Emerging," in Paul K. Davis, ed., *New Challenges for Defense Planning: Rethinking How Much Is Enough,* Santa Monica, Calif.: RAND Corporation, Chapter 2, 1994a, pp. 15–27. As of April 3, 2008:
http://www.rand.org/pubs/monograph_reports/MR400/

———, "Institutionalizing Planning for Adaptiveness," in Paul K. Davis, ed., *New Challenges for Defense Planning: Rethinking How Much Is Enough,* Santa Monica, Calif.: RAND Corporation, Chapter 4, 1994b, pp. 51–73. As of April 3, 2008:
http://www.rand.org/pubs/monograph_reports/MR400/

———, ed., *New Challenges for Defense Planning: Rethinking How Much Is Enough,* Santa Monica, Calif.: RAND Corporation, 1994c. As of April 3, 2008:
http://www.rand.org/pubs/monograph_reports/MR400/

———, "Protecting the Great Transition," in Paul K. Davis, ed., *New Challenges for Defense Planning: Rethinking How Much Is Enough,* Santa Monica, Calif.: RAND Corporation, Chapter 6, 1994d. As of April 3, 2008:
http://www.rand.org/pubs/monograph_reports/MR400/

———, *Analytic Architecture for Capabilities-Based Planning, Mission-System Analysis, and Transformation,* Santa Monica, Calif.: RAND Corporation, 2002. As of April 3, 2008:
http://www.rand.org/pubs/monograph_reports/MR1513/

Davis, Paul K., and Lou Finch, *Defense Planning for the Post–Cold War Era: Giving Meaning to Flexibility, Adaptiveness, and Robustness of Capability,* Santa Monica, Calif.: RAND Corporation, 1993. As of April 3, 2008:
http://www.rand.org/pubs/monograph_reports/MR322/

Davis, Paul K., David C. Gompert, and Richard L. Kugler, *Adaptiveness in National Defense: The Basis of a New Framework,* Santa Monica, Calif.: RAND

Corporation, IP-155, 1996. As of April 3, 2008:
http://www.rand.org/pubs/issue_papers/IP155/

Davis, Paul K., and James P. Kahan, *Theory and Methods for Supporting High Level Decisionmaking,* Santa Monica, Calif.: RAND Corporation, TR-422-AF, 2007. As of April 3, 2008:
http://www.rand.org/pubs/technical_reports/TR422/

Davis, Paul K., Richard L. Kugler, and Richard Hillestad, *Strategic Issues and Options for the Quadrennial Defense Review (QDR),* Santa Monica, Calif.: RAND Corporation, DB-201-OSD, 1997. As of April 3, 2008:
http://www.rand.org/pubs/documented_briefings/DB201/

Davis, Paul K., Jonathan Kulick, and Michael Egner, *Implications of Modern Decision Science for Military Decision-Support Systems,* Santa Monica, Calif.: RAND Corporation, 2005. As of April 3, 2008:
http://www.rand.org/pubs/monographs/MG360/

Davis, Paul K., Russell D. Shaver, and Justin Beck, *Portfolio-Analysis Methods for Assessing Capability Options,* Santa Monica, Calif.: RAND Corporation, 2008. As of April 3, 2008:
http://www.rand.org/pubs/monographs/MG662/

Davis, Paul K., Russell D. Shaver, Gaga Gvineria, and Justin Beck, *Finding Candidate Options for Investment: From Building Blocks to Composite Options and Preliminary Screening,* Santa Monica, Calif.: RAND Corporation, TR-501-OSD, 2007. As of April 3, 2008:
http://www.rand.org/pubs/technical_reports/TR501/

Davis, Steven J., Kevin M. Murphy, and Robert H. Topel, "War in Iraq Versus Containment: Weighing the Costs," March 20, 2003. As of June 4, 2008:
http://gsbwww.uchicago.edu/fac/steven.davis/research/

———, "War in Iraq Versus Containment," February 15, 2006. As of June 4, 2006:
http://gswww.uchicago.edu/fac/steven.davis/research/

Defense Science Board, *Strategic Technology Planning, Volume III of 2006 Summer Study on 21st Century Strategic Technology Vectors,* Washington, D.C.: Office of the Under Secretary of Defense for Acquisition, Technology, and Logistics, 2007.

Dewar, James, *Assumption Based Planning,* London, UK: Cambridge, 2003.

Dreyer, Paul, and Paul K. Davis, *A Portfolio-Analysis Tool for Missile Defense (PAT-MD): Methodology and User's Manual,* Santa Monica, Calif.: RAND Corporation, TR-262-MDA, 2005. As of April 3, 2008:
http://www.rand.org/pubs/technical_reports/TR262/

Dudley, Susan E., and Sharon L. Hays, *Updated Principles for Risk Analysis,* Washington, D.C.: Executive Office of the President, Office of Management and Budget and Office of Science and Technology Policy, 2007.

Elton, Edwin J., Martin J. Gruber, Stephen J. Brown, and William N. Goetzmann, *Modern Portfolio Theory and Investment Analysis,* New York: John Wiley & Sons, 2006.

Enthoven, Alain, and K. Wayne Smith, *How Much Is Enough? Shaping the Defense Program, 1961–1969,* Santa Monica, Calif.: RAND Corporation, 1971; reissued in 2005.

Fisher, Gene Harvey, *Cost Considerations in Systems Analysis,* New York: Elsevier, 1971.

Fitzsimmons, Michael, "Whither Capabilities-Based Planning," *Joint Forces Quarterly,* Vol. 44, No. 1, 2007.

Galbraith, Jay R., *Designing the Customer-Centric Organization: A Guide to Strategy, Structure, and Process,* San Francisco, Calif.: Jossey-Bass, a Wiley Imprint, 2005.

Garamone, Jim, "Gates Calls for 92,000 More Soldiers, Marines," American Forces Press Service, January 11, 2007. As of June 12, 2008: http://www.defenselink.mil/news/newsarticle.aspx?id=2651

Government Accountability Office, *Best Practices: An Integrated Portfolio Management Approach to Weapon System Investments Could Improve DoD's Acquisition Outcomes,* Washington, D.C.: GAO-07-388, 2007.

Gilmore, J. Michael, *CBO Testimony: The Navy's DD(X) Destroyer Program,* Washington, D.C.: Congressional Budget Office, Congress of the United States, 2005.

Goeller, Bruce F., Allan F. Abrahamse, J. H. Bigelow, J. G. Bolten, David M. De Ferranti, James C. DeHaven, T. F. Kirkwood, and R. L. Petruschell, *Protecting an Estuary from Floods. Vol. I, Summary Report: A Policy Analysis of the Oosterschelde,* Santa Monica, Calif.: RAND Corporation, 1977. As of June 4, 2008: http://www.rand.org/pubs/reports/R2121.1/

Gompert, David C., John Gordon, Adam Grissom, Dave Frelinger, Seth G. Jones, Martin C. Libicki, Edward O'Connell, Brooke K. Stearns, and Robert Edwards Hunter, *War by Other Means: Building Complete and Balanced Capabilities for Counterinsurgency: RAND Counterinsurgency Study, Final Report,* Santa Monica, Calif.: RAND Corporation, 2008. As of April 3, 2008: http://www.rand.org/pubs/monographs/MG595.2/

Gompert, David, Paul K. Davis, Stuart Johnson, and Duncan Long, *Analysis of Strategy and Strategies of Analysis,* Santa Monica, Calif.: RAND Corporation, 2008. As of August 12, 2008: http://www.rand.org/pubs/monographs/MG718/

Greer, W. L., *Exploration of Potential Future Fleet Architectures,* Alexandria, Va.: Institute for Defense Analyses, 2005.

Grissom, Adam, and David Ochmanek, "Train, Equip, Advise, Assist: The USAF and the Indirect Approach to Countering Terrorist Groups Abroad," Santa Monica, Calif.: RAND Corporation, 2008. Not available to the general public.

Hagstrom, Robert G., *The Warren Buffett Portfolio: Mastering the Power of the Focused Investment Strategy,* New York: John Wiley & Sons, 1999.

Haimes, Yacov, *Risk Modeling, Assessment, and Management,* New York: John Wiley & Sons, 1998.

Hillestad, Richard, and Paul K. Davis, *Resource Allocation for the New Defense Strategy: The DynaRank Decision Support System,* Santa Monica, Calif.: RAND Corporation, 1998. As of April 3, 2008:
http://www.rand.org/pubs/monograph_reports/MR996/

Hitch, Charles J., and Roland N. McKean, *The Economics of Defense in the Nuclear Age,* New York: Scribner's Sons, 1965. Published originally by the RAND Corporation and Harvard University Press, 1960. As of June 4, 2008:
http://www.rand.org/pubs/reports/R346/

Johnson, Joseph Andrew, *Cost Estimates for Desert Shield/Desert Storm: A Budgetary Analysis,* Monterey, Calif.: U.S. Naval Postgraduate School, 1991.

Johnson, Stuart, and Arthur K. Cebrowski, *Alternative Fleet Architecture Design,* Washington, D.C.: Center for Technology and National Security Policy, National Defense University, 2005.

Johnson, Stuart, Martin Libicki, and Gregory F. Treverton, eds., *New Challenges, New Tools for Defense Decisionmaking,* Santa Monica, Calif.: RAND Corporation, 2003. As of April 3, 2008:
http://www.rand.org/pubs/monograph_reports/MR1576/

Joint Defense Capabilities Study Team, *Joint Defense Capabilities Study: Improving DoD Strategic Planning, Resourcing and Execution to Satisfy Joint Capabilities (the "Aldridge Report"),* Washington, D.C.: Department of Defense, 2004.

Kahn, Herman, and Irwin Mann, *Techniques of Systems Analysis,* Santa Monica, Calif.: RAND Corporation, 1957. As of April 3, 2008:
http://www.rand.org/pubs/research_memoranda/RM1829-1/

Kaplan, Robert S., and David P. Norton, *The Balanced Scorecard: Translating Strategy into Action,* Cambridge, Mass.: Harvard Business School Press, 1996.

Kohyama, Hiroyuki, and Allison Quick, *Accrual Accounting in Federal Budgeting: Retirement Benefits for Government Workers,* Cambridge, Mass.: Harvard Law School, Briefing Paper No. 25, 2006.

Lempert, Robert J., "A New Decision Science for Complex Systems," *Proceedings of the National Academy of Sciences Colloquium,* Vol. 99, Suppl. 3, May 14, 2002.

Lempert, Robert J., David G. Groves, Steven W. Popper, and Steve C. Bankes, "A General Analytic Method for Generating Robust Strategies and Narrative Scenarios," *Management Science,* Vol. 33, No. 2, April 2006.

Lempert, Robert J., Steven W. Popper, and Steven C. Bankes, *Shaping the Next One Hundred Years: New Methods for Quantitative Long-Term Policy Analysis,* Santa Monica, Calif.: RAND Corporation, 2003. As of April 3, 2008: http://www.rand.org/pubs/monograph_reports/MR1626/

Light, Paul C., *The Four Pillars of High Performance,* New York: McGraw-Hill, 2004.

Madansky, Albert, "Uncertainty," in Edward S. Quade and Wayne I. Boucher, eds., *Systems Analysis and Policy Planning: Applications in Defense,* Santa Monica, Calif.: RAND Corporation, 1968, pp. 81–96.

Markowitz, Harry M., "Portfolio Selection," *Journal of Finance,* Vol. 7, No. 1, 1952, pp. 77–91.

Martel, J. M., N. T. Khoury, and M. Bergeron, "An Application of Multicriteria Approach to Portfolio Comparisons," *Journal of the Operational Research Society,* Vol. 39, No. 7, 1988, pp. 617–628.

Moore, Richard M., John Stillion, Clarence R. Anderegg, Susan R. Bohandy, Michael Boito, James Bonomo, Michel Scott Brown, Katherine M. Calef, James S. Chow, Leo C. Cloutier, Toby Edison, Scot Eisenhard, Dave Frelinger, Mark David Gabriele, Roy O. Gates, Elham Ghashghai, Jonathan Gary Grossman, Jeff Hagen, Gail Halverson, Thomas J. Herbert, Walter Hobbs, Alex Hou, Phyllis Kantar, Michael Kennedy, Beth E. Lachman, Robert S. Leonard, David M. Matonick, D. Norton, David T. Orletsky, Harold Scott Perdue, Kevin L. Pollpeter, Bob Preston, Randall Steeb, Carl W. Stephens, F. S. Timson, Michael S. Tseng, and Barry Wilson, *Next-Generation Gunship Analysis of Alternatives, Vol. 1: Final Report,* Santa Monica, Calif.: RAND Corporation, 2004. Not available to the general public.

Morgan, M. Granger, and Max Henrion, *Uncertainty: A Guide to Dealing with Uncertainty in Quantitative Risk and Policy Analysis,* New York: Cambridge, 1990.

National Academy of Sciences, *Conventional Prompt Global Strike Capability,* Washington, D.C.: National Academies Press, forthcoming.

National Research Council, *Naval Analytical Capabilities: Improving Capabilities-Based Planning,* Washington, D.C.: National Academies Press, 2005.

———, *Defense Modeling, Simulation, and Analysis: Meeting the Challenge,* Washington, D.C.: National Academies Press, 2006.

———, *Scientific Review of the Proposed Risk Assessment Bulletin from the Office of Management and Budget,* Washington, D.C.: National Academies Press, 2007.

Nordhaus, William D., *War with Iraq: Costs, Consequences, and Alternatives,* Cambridge, Mass.: American Academy of Arts & Sciences, 2002.

O'Rourke, Ronald, *Potential Navy Force Structure and Shipbuilding Plans: Background and Issues for Congress,* Washington, D.C.: Congressional Research Service, 2004.

———, *Navy Littoral Combat Ship (LCS) Program: Oversight Issues and Options for Congress,* Washington, D.C.: Congressional Research Service, 2007.

Office of Management and Budget, "Circular A-4." As of June 4, 2008: http://www.whitehouse.gov/omb/circulars/a004/a-4.html

———, "Discount Rates for Cost-Effectiveness, Lease Purchase, and Related Analyses." As of June 4, 2008: http://www.whitehouse.gov/omb/circulars/a094/a94_appx-c.html

Orszag, Peter, *Estimated Costs of U.S. Operations in Iraq and Afghanistan and Other Activities Related to the War on Terrorism,* Statement of Peter Orszag Before the Committee on the Budget, U.S. House of Representatives, October 24, 2007, Washington, D.C.: Congressional Budget Office, Congress of the United States, 2007.

Pace, General Peter, *National Military Strategy of the United States,* Washington, D.C.: Joint Chiefs of Staff, 2004.

Quade, Edward S., and Wayne I. Boucher, eds., *Systems Analysis and Policy Planning: Applications in Defense,* New York: Elsevier Science Publishers, 1968.

Quade, Edward S., and Grace M. Carter, eds., *Analysis for Public Decisions,* 3d ed, New York: North-Holland, 1989.

Rumsfeld, Donald, *Report of the Quadrennial Defense Review,* Washington, D.C: Department of Defense, 2001.

———, *National Defense Strategy of the United States of America,* Washington, D.C.: Department of Defense, 2005a.

———, *National Defense Strategy of the United States,* Washington, D.C.: Department of Defense, 2005b.

———, *Quadrennial Defense Review Report,* Washington, D.C.: Department of Defense, 2006.

Sirak, Michael, "Senior Air Force General to Skeptics: We Can Field a New Bomber in 2018," *Defense Daily,* 2007.

"SSGN: SSGN Guided Missile Submarines," Department of the Navy, Research, Development and Acquisition, no date. As of June 13, 2008: http://acquisition.navy.mil/programs/submarines/ssgn

Sunstein, Cass R., and Arden Rowell, *On Discounting Regulatory Benefits: Risk, Money, and Intergenerational Equity,* AEI-Brookings Joint Center for Regulatory Studies, Working Paper 05-08, 2005. As of July 15, 2008: http://papers.ssrn.com/sol3/papers.cfm?abstract_id=756832

Swisher, Pete, and Gregory W. Kasten, "Post-Modern Portfolio Theory," Journal of Financial Planning, Vol. 18, No. 9, 2005, pp. 74–85.

Technical Cooperation Program, *Guide to Capability-Based Planning*, Alexandria, Va.: The Technical Cooperation Program, TR-JSA-TP3-2-2004, 2004.

U.S. Air Force, *Committee Staff Procurement Backup Book, FY2008/2009 Budget Estimates, Volume I,* February 2007, p. 4/45. As of June 13, 2008: http://www.saffm.hq.af.mil/shared/media/document/AFD-070212-005.pdf

U.S. Congress, *Goldwater-Nichols Department of Defense Reorganization Act of 1986, Public Law 99-433,* Washington, D.C.: United States Congress, 1986.

Woolf, Amy F., *Conventional Warheads for Long-Range Ballistic Missiles: Background and Issues for Congress,* Washington, D.C.: Congressional Research Service, 2007.